VALUE CHAIN FINANCE

VALUE CHAIN FINANCE

BEYOND MICROFINANCE FOR RURAL ENTREPRENEURS

Royal Tropical Institute

IIRR

This publication is jointly produced by:

Royal Tropical Institute

Royal Tropical Institute (KIT), PO Box 95001, 1090 HA, Amsterdam, The Netherlands
development@kit.nl, www.kit.nl

IIRR

International Institute of Rural Reconstruction (IIRR), Africa Regional Centre, PO Box 66873, Nairobi, Kenya
admin@iirr.org, www.iirr.org

The writeshop to produce this book was made possible with the financial support from:

- **DGIS**, Netherlands Directorate-General of Development Cooperation
- **Ford Foundation**
- **Hivos**, Humanist Institute for Development Cooperation
- **ICCO**, Interchurch Organisation for Development Co-operation
- **Terrafina Microfinance**
- **Triodos Sustainable Trade Fund**

FORDFOUNDATION

people unlimited Hivos

icco Partner to enterprising people

Terrafina Microfinance

Triodos Sustainable Trade Fund

KIT and IIRR are grateful to all persons and organizations who shared their knowledge and experiences in the writeshop where this book was drafted. Sincere thanks to the staff of IIRR's Africa regional office, who provided an excellent facilitation of the writeshop. A special word of gratitude to the farmers' organizations, development organizations, companies and financial institutions in Africa, Asia and Latin America whose experiences are reflected in this book. Many thanks to the overall editor for an excellent job in editing and layout of the book, and to the proofreader for proofreading. Page xvi lists the names of indviduals who contributed to the book.

Coordination: Lucian Peppelenbos

ISBN: 978-94-6022-055-5

Order from: publications@iirr.org (for Africa) and publishers@kit.nl, www.kitpublishers.nl (rest of the world)

Correct citation: KIT and IIRR. 2010. *Value chain finance: Beyond microfinance tor rural entrepreneurs*. Royal Tropical Institute, Amsterdam; and International Institute of Rural Reconstruction, Nairobi.

FSC Mixed Sources
Productgroep uit goed beheerde bossen en andere gecontroleerde bronnen
www.fsc.org Cert no. SGS-COC-006507
© 1996 Forest Stewardship Council

Contents

Boxes

Figures

Tables

Foreword

VALUE CHAIN DEVELOPMENT IS a growing approach worldwide to increase incomes of small producers and the economically active poor. Access to adequate and timely financial services for all actors in the chain has proven a key element for success. This implies that not only large producers and traders but also small producers need access to appropriate financial services to make optimal use of value addition and income generation. Such finance is, however, not always available, and chain actors working in agricultural and rural value chains frequently complain about a lack of access to financial services.

Traditionally, most regular banks and microfinance institutions have avoided rural finance since this is perceived as risky and costly, with cash flow requirements that are irregular and difficult to manage. Banks shy away from the high transaction costs and risks related to agriculture such as crop failure, diseases and market fluctuations. Also the lack of physical collateral is a restriction, and the risk of political interference that can damage the repayment behaviour of the rural clientele is high. Most microfinance institutions opt for high-density urban or peri-urban areas where they serve their clients in standardized – often group-based – systems, usually unfit for the needs of small farmers. In answer to this gap in finance, there has been a tendency for companies worldwide to patronize a certain chain as a whole and directly finance producers or traders. While this may be a good short-term solution to a burning problem, one may also question whether this will provide enough perspective for sustainability and scaling up of chain interventions needed to reach out to many millions of small rural producers and processors worldwide. Also one can question whether such financing mechanisms will contribute to the desired empowerment of small rural producers and processors, or rather increase their dependency on larger chain actors.

In this book the link between chain actors and financial institutions is described as a means to deepen financial services for value chains. Value chain finance aims to address perceived constraints and risks by providing innovative ways of delivering financial services to rural producers and agribusinesses. Value chain finance means linking financial institutions to the value chain, offering financial services to support the product flow, and building on the established relationships in the chain. It means that the product flow in the value chain is used as a carrier to provide financial services. This way of financing can spread risk among the financial institutions and chain actors and provides alternatives to traditional collateral requirements. It provides tremendous potential for unleashing capital, scaling up and sustaining chain prospects, but it needs to be managed and organized well.

Value chain finance as defined in this book needs to build on trust and strong relationships between chain actors and financial service providers. Parties need to know and understand each other, and this often requires a change of mindset, particularly with the banks but also with the microfinance institutions, producers and companies. All cases in the book underline the need for financial institutions to have a deep understanding of the realities in the chain and of the need for timely and flexible finance. Likewise, the chain actors need to understand the business realities and mentalities of banks and microfinance institutions looking at risk mitigation and cost coverage for their services.

Innovative financial products and services are needed such as insurance, overdrafts, factoring and leasing models, as well as investment loans, guarantees and venture capital. Such a variety of sizes and types of financial products can be offered only through a combination of financial service providers. This requires building vertical linkages in the financial sector. For example, microfinance institutions can link with producer organizations to provide small input loans to producers, while banks simultaneously provide an investment loan to a processing company in the chain. The microfinance institutions and banks need to link up to align their services in this chain for potential overdraft facilities to small traders.

Organizations such as Hivos, ICCO, Terrafina Microfinance and Triodos Bank can be instrumental in horizontal and vertical relationship building between producers, companies and financial service providers. These organizations are promoting value chain finance as part of their policies to develop rural entrepreneurship, and can develop a variety of support mechanisms to financial institutions and chain actors. Such mechanisms are indispensable for increasing the numbers of small rural producers and processors that make use of value chain finance.

This book is the third in a series on value chain development by the Royal Tropical Institute (KIT) and the International Institute of Rural Reconstruction (IIRR) and is the logical follow up to the two other books. One is called *Chain empowerment: Supporting African farmers to develop markets* (2006) and the other is titled *Trading up: Building cooperation between farmers and traders in Africa* (2007). This book is the first to draw on worldwide experiences.

This publication will add to the growing literature on value chain development and value chain finance. We hope that the publication contributes to ongoing exchange and mutual learning and will increase the effectiveness of the work on poverty reduction of our organizations. We also hope that the book will stimulate bankers and government institutions to support value chain finance in a bid to create sustainable local and international markets. This book is a result of exchange by practitioners and academics using the particular method of a "writeshop" (jointly writing a book in a workshop). As a result it contains an easy-to-read analysis of many excellent examples from practice that convince us that, together, development agents can increase their relevance for many millions of rural poor.

FORD FOUNDATION, HIVOS, ICCO, TERRAFINA MICROFINANCE, TRIODOS BANK

Preface

IN 2006, THE ROYAL Tropical Institute (KIT) and the International Institute of Rural Reconstruction (IIRR) published a best-seller on empowering African smallholder farmers in value chains. The book was reprinted in 2007, and in 2008 we published a second book on building cooperation between farmers and traders in Africa. This book was also well received, and the first printing of 1000 copies was sold out before the end of the year.

The writeshop approach used by IIRR and KIT proved able to result in interesting reading for practitioners who recognize themselves in the real-life experiences documented in these books. It is a simple phenomenon: audiences appreciate a movie more if they can sympathize with the hero. The people who help us write and publish these books are all champions in their fields and share their best practices to the inspiration of others.

When considering what issue should feature next in our series, we came across what experts call the "finance gap". Farmers can be empowered to participate in value chain governance. Traders can build strong relations with farmers to achieve value chain efficiency. But if there is no finance to make the chain work, business comes to an abrupt stop. Successful business needs trade finance and loans for making investments. No normal business venture can finance this with its own resources. Entrepreneurs need financial services from banks and other financial agents to keep operating and growing their businesses.

In recent years the microfinance industry has mushroomed to develop into a big business catering for the needs of previously "unbankable" micro-entrepreneurs around the world. Yet small- and medium-sized enterprises still experience major problems in borrowing money for their operations and investments, especially in the countryside. This is the finance gap that constrains business development in Africa, Asia and Latin America.

Business people in low-income countries encounter major obstacles for obtaining loans. Banks and related financial institutions have built high walls for small and medium-sized businesses. Their requirements and procedures seem to be designed to keep business people out of the bank. Often, applying for a bank credit is more a matter of political networking than doing good business.

But these enterprises employ lots of people, are able to invest in product development, and are capable of exploring new markets. There is plenty of evidence that small and medium-sized enterprises are the real backbone of developing economies. Not providing these small engines of development with capital to operate is a tragedy in disguise.

Fortunately there are heroes in this story. There are people and financial institutions in Africa, Asia and Latin America who look for solutions that cross the boundaries of the conventional. They understand the problem, but also the opportunities offered by small and medium enterprises. They are developing tools and mechanisms to provide them with capital. And they are succeeding in promoting good business and achieving sound returns on their investment.

This book is about their experiences. The authors of this book are, again, practitioners themselves, who tell their own stories, which we have carefully put into context. I am sure that you as a reader of this book will sympathize with them and admire their novel ways of providing financial services.

On behalf of KIT and IIRR, I would like to thank all the contributors for their willingness to explain how their businesses and how their financial models work. There are still relatively few such initiatives around the world, and it is high time that the pioneering approaches were copied, adapted and expanded to the benefit of millions of other farmers and businesses.

I would also like to thank our donors and partners who made this book possible. Their support is vital to KIT's and IIRR's efforts to promote better approaches to value chain development.

Finally, my personal thanks to the team of specialists, facilitators, editors, artists and other staff, within KIT and IIRR and outside, who have made this book possible.

Bart de Steenhuijsen Piters

Area Leader Sustainable Economic Development

KIT - Royal Tropical Institute

List of contributors

For further information and contact details, see page 246.

* Did not attend writeshop

Participants

Bolivia

Sartawi Foundation
Marcio Oblitas Fernández

Save the Children
Navil Agramont Akiyama

Pro-Rural
María Renée Bejarano Vega

Ethiopia

Facilitators for Change Ethiopia (FCE)
Tarekegn Garomsa

Harbu Microfinance Institution Share Company
Tesfaye Befekadu

India

ACDI/VOCA
Amit Kumar Singh

BASIX Group
Mendu Srinivasulu

Catalyst Group
N Raghunathan*
Nagabhushana Clospet Satyanarayana Rao

Foundation for MSME Clusters
Sangeeta Agasty

Reliance Dairy Foods
Beeranna Palyam

South Indian Federation of Fishermen Societies (SIFFS)
Ephrem Soosai
V. Vivekanandan*

Kenya

Kenya Gatsby Trust
Billy O Mathu

K-Rep Development Agency
Paul Maundu Mwilu

Kwieria Tangi Women Group
Joseph Njoroge Ndirangu

Nakuru Region Inter-Diocesan Christian Community Services (NRIDCCS)
Joseph Muriithi Ndegwa

Netherlands

HIVOS
Harry Clemens

Royal Tropical Institute (KIT)
Maurits de Koning
Anna Laven
Lucian Peppelenbos

Nicaragua

Fondo de Desarrollo Local
Alexandra Schütze
Manuel Antonio Bermúdez Vado

Peru

Fomento de la Vida (Fovida)
Luisa Santur Alberca

Rwanda

Caisse des Affaires Financières Isonga (CAF Isonga)
Joseph Kabundi

Cooperative COPRORIZ
Callixte Niyonsaba

SNV Netherlands Development Organisation
Straton Habyalimana

United Kingdom

Independent consultant
Malcolm Harper

Tanzania

BioRe Tanzania
Niranjan Pattni

Writeshop and production

Coordination
Anna Laven
Lucian Peppelenbos

Resource persons
Malcolm Harper
Maurits de Koning
Anna Laven
Lucian Peppelenbos

Editing
Paul Karaimu
Paul Mundy (overall editing)
Aileen Ogolla
Sarah Simpson

Facilitation
Isaac Bekalo Bateno
Janet Nyaoro

Graphics, layout
Paul Mundy

Illustrations
Nyotumba Bonaventure
Alfred Ombati

Logistics
Oliver Mundy
Julia Wagemann

Proofreading
Fred Zaal

1

Linking rural entrepreneurs to financial services

CONRADO GROWS POTATOES ON a plot of land in the Valles region in Bolivia. He is proud of his crop: the tubers are good quality, and his uncle, a trader, pays him a good price for them. Although business is going quite well, Conrado has a problem: every day he has to walk 2 hours to get to his farm. If he had a bicycle, he could spend this time instead grading and packaging the potatoes, which would fetch him a higher price. His uncle cannot afford to lend him the money – he has to pay staff and the costs of storing the potatoes. Conrado went to a local bank, but they refused him a loan: Conrado does not own the land he works on; without collateral, the bank cannot help him.

Gonzalez, a bank employee, knows the Valles region well: his father used to grow potatoes there, and made just enough money to put his children through school. Gonzalez started working at the bank with a dream: he thought his job would let him help poor people improve their businesses. But even though he knew that an ambitious young man like Conrado would be able to pay back his loan, he could not help him. He asked his boss if an exception could be made. His boss said no: the bank's rules prohibited making loans without collateral.

* * *

This story illustrates a mismatch between a would-be borrower and a potential lender. Even knowing that a loan to the farmer would generate benefits for both parties, the bank cannot enter this kind of business. Generally there is a problem in how financial services suit the demands and circumstances of rural clients. This book is about how rural entrepreneurs can be better linked to providers of financial services.

Only 20 years ago, giving credit to poor people was seen as foolish. Since then, improvements have been made. Poor households increasingly have access to small loans from microfinance institutions, and large agribusinesses can get credit through commercial banks. But many rural producers and small- and medium-sized agribusinesses remain under-served. Their financial needs are generally too large for microfinance, but too small for commercial banks. This gap hampers growth and limits agricultural development. It is also a loss to the financial sector, which ignores millions of potential rural clients.

Financial service providers generally regard the rural sector as too risky and as involving high transaction costs. They doubt the ability and willingness of rural entrepreneurs to repay their debts. Potential lenders may also see high risks because they lack understanding of the rural sector, and have no way to evaluate the risks in agricultural value chains (Quirós 2006).

In this book we explore **value chain finance** as a way to deliver more, and better, financial services to farmers and rural entrepreneurs. Realizing the benefits of value chain finance requires a shift in thinking on the risks in financing agriculture, and on how to mitigate risks through new approaches to cooperation. Such a shift can neither be made overnight, nor by every financial institution. But by sharing experiences with value chain finance, this book aims to pave the way for financial institutions to increase the reach of their services to small-scale farmers and other rural entrepreneurs. We also hope that rural entrepreneurs and development organizations will be able to draw lessons and inspiration on ways to promote business and development in rural areas.

Parts of this book

The remainder of this book is divided into seven chapters.

Chapter 2, Finance for rural entrepreneurs, focuses on the finance gap for rural entrepreneurs, and introduces the role of value chain finance in delivering services to the poor.

Chapter 3, A value chain finance framework, takes a closer look at how value chain finance works, and presents a framework that helps us analyse the cases in the following chapters.

Chapter 4, Crafting new chains, presents four cases that involved establishing new value chains: potatoes in Peru, milk in Bolivia and India, and chilli in Kenya. In these cases, financial services were part of an integrated approach to building a value chain.

Chapter 5, Improving chain liquidity, explores four cases where financial agents linked with existing chains and provided new services to build on and improve the chain flows. These chains cover cotton in Tanzania, sal leaf in India, tea in Kenya, and rice in Rwanda.

Chapter 6, Unleashing investments in the chain, focuses on how chain actors made medium-term investments and how financial agents played an enabling role in this process. The five cases in this chapter cover honey in Kenya, soybean in Ethiopia, coffee in Nicaragua, quinoa in Bolivia, and fish in India.

Chapter 7, Synthesis, looks at the cases through a wider lens. It draws lessons from the cases and offers some ideas on ways to use value chain finance to promote rural businesses in ways that benefit the poor.

Chapter 8, Resources, lists organizations, websites and publications that focus on value chain finance. It also gives the contact details of the people who contributed to the book.

How this book was produced

This book is part of wider efforts to generate practical knowledge for organizations that work on the development of markets, value chains and financial services in Africa. In 2006 KIT, IIRR and Faida MaLi jointly produced a book, *Chain empowerment: Supporting African farmers to develop markets* that reflected the experiences and views of many organizations assisting African farmers in value chain development. The book was well received as many organizations found it useful as a guide in their programmes and activities with the farmers.

In 2008 KIT and IIRR produced a second book focusing on the role of traders in value chains, *Trading up: Building cooperation between farmers and traders in Africa.* Many organizations working on value chain development are unfamiliar with how to deal with traders. They therefore encourage the farmers to do the marketing themselves, which in many cases has yielded disappointing results. This book showed that cooperation with traders is indeed possible, and can bring benefits for all.

This third book turns the focus to the role of finance in value chains. Financial markets in developing countries largely fail to serve the need of small- and medium-sized enterprises, particularly in rural areas (the "missing middle"). Among the many current innovations, value chain finance is a promising one. This book presents some experiences in value chain finance, assesses the preconditions, opportunities and limitations of this approach, and develops experience-based strategies and guidelines for future policy and practice.

For this book, KIT played an overall technical coordination role, developed a conceptual framework and raised funds, while IIRR advised on the process, assembled the technical team, facilitated the writeshop and organized the logistics. KIT and the editor managed the editing and layout, and KIT Publishers was responsible for printing.

The production of this book was supported by ICCO, Triodos Bank, Terrafina Microfinance, Hivos and the Ford Foundation. Hivos staff also participated in the writeshop.

The bulk of this book was produced through an intensive participatory "writeshop", held from 18 to 27 February 2009 in Nairobi, Kenya. Prior to the writeshop, potential cases were identified through an open call for cases sent to the partner networks of KIT, IIRR and the donor organizations. On the basis of one-page abstracts, 13 cases from Africa, Latin America and Asia were selected. The authors were sent guidelines on how to write their case along with a sample to use as a model, and were invited to participate in the writeshop.

Each contributor brought to the writeshop a draft manuscript describing interventions by an external financial institution providing services to chain actors. Each case focused on how value chain finance contributed to smoothing of the chain and competitiveness. They were asked to bring with them to the writeshop other printed materials and photographs relevant to their case.

The 37 participants included managers and staff of financial institutions, business development services, private companies and cooperatives, as well as farmers, development professionals, researchers, facilitators, artists and editors. They are listed on page xvi, and their contact details are given on page 246.

The writeshop process

The writeshop began with an introduction to the process to be used and the framework that underlies this book (described further in Chapter 3). One of the cases was presented as an example, making use of the framework. On the second day the participants worked in teams, together with a resource person (individual specialists in value chain finance) and an editor, and prepared a presentation of their case, following the first example.

The authors then presented their cases in turn to the plenary. After each presentation, the participants had an opportunity to ask questions, make comments, and critique the manuscript. The author, editor and resource person allocated to that case took notes. With their assistance, the authors then rewrote their drafts, adding details about the situation before and after the financial initiative described in the case. An artist drew illustrations to depict the case. Teams of participants also used labels, marker pens and flip-chart paper to depict the flowcharts of their value chains (these were later condensed into the graphics included in this book).

The authors then in turn presented their revised manuscripts to two sub-plenary groups. The other participants again commented on and critiqued each case, and the author, resource person and editor again took notes, then incorporated the corrections into a third draft.

Also during the writeshop, participants took part in two brainstorming and discussion sessions to generate ideas for the Analysis section of the book (Chapter 7). In the first, groups of participants discussed the themes of "crafting new value chains", "expanding chain liquidity" and "unleashing investments in the chain". The second session dealt with cross-cutting issues: "building and destroying triangles", "opportunities, benefits and limitations", and "institutional environment and governance".

One day of the writeshop consisted of a forum for stakeholders and policymakers, where financial institutions and policymakers shared their experiences with innovative ways of financing chain actors. A second day was devoted to a field visit to a milk cooperative and a small milk farmer near the writeshop venue.

After the writeshop, only minor editing was necessary to put the case manuscripts (which form the bulk of the book, Chapters 4–6) into their final form. The KIT coordinators wrote the introductory and analysis chapters (Chapter 1–3 and 7–8) based on the ideas developed during the brainstorming sessions. The overall editor in collaboration with KIT and IIRR was responsible for finalizing the book.

Throughout the writeshop process, the initial manuscripts were revised substantially or were completely rewritten. The individual participants remain the main authors of their cases; their names are printed at the end of each case.

Writeshop advantages

The sequence described above is an adaptation of the writeshop approach pioneered by IIRR at its headquarters in the Philippines. IIRR–Africa has used this approach to produce extension and information materials on a wide range of subjects. Writeshops have several advantages over conventional methods of producing a publication. They speed up the production process, taking full advantage of the participants' range of expertise. The process of writing, getting comments, revising and illustrating takes place at the same time, considerably shortening the often-difficult process of writing, editing and publishing. A large number of participants contribute to each topic, generating insights and significant amounts of new information. Working with the editors enables the manuscripts to be revised, simplified and put into a common format. In effect, the writeshop provides an opportunity for technical peer review by a large number of reviewers, as well as pre-testing for understandability and field relevance by a group of the intended readers.

In addition, writeshops bring together a large number of people from various institutions, countries and walks of life, each with different perspectives and expertise. They are an excellent training and networking opportunity, with individuals learning about each other's work and exchanging ideas and experiences that will be of value for them when they return home. It is hoped that the relationships and networks forged during the writeshop will continue long into the future.

2

Finance for rural entrepreneurs

AGRICULTURE IS THE BACKBONE of the economy in developing countries. Millions of micro-entrepreneurs – farmers, processors, traders, transporters, input suppliers – run their businesses in difficult circumstances. Roads are potholed, distances are long, market prices are often unknown, inputs may not be available, electricity is unreliable – these entrepreneurs need to deal with many challenges on a daily basis. All this points to enormous entrepreneurial potential. But most of these entrepreneurs are asset-poor and their management is informal, so financial institutions are reluctant to finance them.

The rural finance gap

Without finance, farmers may not be able to buy good seeds, hire workers, or invest in equipment. For traders, a lack of finance may mean that they cannot pay cash when they take delivery of the crops – so the farmers may sell their crops elsewhere. For small-scale processors, a lack of finance may mean they cannot expand their operations.

Private financial institutions have tended to regard such micro-entrepreneurs as unbankable. Banks did not think they were creditworthy – micro-entrepreneurs have no credit histories or collateral to offer; many are illiterate, so cannot fill in the necessary paperwork. For bankers it is easier and more lucrative to provide a handful of large loans to well-established businesses, rather than lots of small loans to such micro-entrepreneurs (Yunus 2007: 47–8).

An additional problem is that these businesses are in the countryside. Agriculture is a risky business. Drought, heavy rain, pests and diseases, unreliable input supplies, lack of storage and cooling facilities, bumpy roads, fluctuating prices, seasonality of many crops: all make the financial outcome of farming unpredictable (Fries and Akin 2004), so most banks are reluctant to finance crops and livestock. They have few staff or branches in the countryside, and distances are large, pushing up transaction costs.

The bad reputation of agricultural credit does not help either. From the 1950s to the late 1980s, public bodies intervened extensively in rural credit markets in developing countries, especially in Africa. Governments and international donors used heavy subsidies to promote rural lending. Credit was cheap, and it often went to the wrong people, at the wrong time, for the wrong purpose. When farmers had difficulty in repaying their loans, or deliberately defaulted on their repayments, no measures were taken against them. Due to low repayment rates and bad allocation, rural credit programmes became permanently dependent on external resources, and were not a viable operation for private banks.

The result is a serious and long-lasting rural finance gap (UNCTAD 2004) that keeps the economic potential of agriculture underused. The perception of agriculture as risky means a loss to the farmers and other entrepreneurs, to the local and national economy, and to the financial sector. It hinders agricultural development and blocks attempts to reduce poverty. Improved financial markets in rural areas would stimulate agricultural and rural growth, leading to economic growth and less poverty (USAID 2005).

With commercial banks reluctant to lend to the rural poor, and public agricultural development banks closed because of bad performance, it took microfinance institutions to prove that the asset-poor are bankable. With the use of new lending techniques, the microfinance industry showed that lending to micro-entrepreneurs is not only feasible, but can even be an attractive market opportunity.

Microfinance: The beginnings

In the 1970s and 1980s, microfinance institutions began providing small loans to poor people. Muhammad Yunus, winner of the Nobel Peace Prize in 2006, pioneered microcredit. He set up a project testing the idea of lending small amounts of money to the poor. The project showed that the poor are very well able to pay back their loans. So in 1983, Yunus created a special bank for this purpose in Bangladesh, called Grameen Bank (which means "village bank").

Grameen Bank was revolutionary: it represented a shift in thinking, challenging the belief that loans cannot be made without collateral. The bank empowered its borrowers to lift themselves out of poverty by removing the institutional barriers that normally prevent poor people from accessing financial services. The bank showed that the poor are entrepreneurs: they repay their loans with interest, using money they earn through their own productive work.

One important innovation underlying the success of Grameen Bank is the use of positive social pressure to create trust and loyalty. The bank provides loans to groups of borrowers: "No one who borrows from Grameen Bank stands alone" (Yunus 2007: 57). Preferably, these groups of borrowers are self-formed by neighbours and friends who meet regularly. These small social networks are embedded in a larger group: a centre where ten to twelve groups come together for weekly meetings. This "community-oriented dynamic" is perceived as one of the cornerstones of Grameen's success and has also been promoted among commercial banks.

Other new techniques also explain the success of microfinance vis-à-vis traditional banking:

- **Regular repayments and savings** Building on techniques used by traditional grassroots saving groups (Box 2.1), Grameen introduced the idea of regular repayments and savings. Borrowers within a group have to repay their loans regularly; they have to repay completely before others can apply for a loan. That creates social pressure within the group and the centre to encourage borrowers to remain faithful to their commitments. This has contributed to the high repayment rates of Grameen Bank; in 2006 this was almost 99% (Yunus 2007: 51). Yunus says this success is partly because the bank looked at the behaviour of people it lent money to. This "people-oriented approach" is illustrated by the preference to give credit to women instead of men; unlike men, women tend not to spend the money on themselves but on their families.
- **Non-financial services** Finance alone cannot reduce poverty. Grameen offers multiple services, such as information technology, scholarships, health and welfare. It also promotes a strong social agenda: every borrower of the bank must commit to this, and is expected to take responsibility for issues such as family planning, education, hygiene and community development.

Despite its success, Grameen Bank also faced difficulties and limitations. Like any business, it has had to adapt over time to serve its customers and their needs more effectively. For example, the bank had to match its services better to its clients' needs. It introduced a wider variety of loan types, as well as pension funds, loan insurance and other financial services. It also had to increase the amount of savings deposits to improve its capital structure and create a reserve. This would enable the bank to become fully self-sufficient.

Worldwide expansion of microfinance

From the pioneering work of Grameen Bank in Bangladesh, microfinance started to conquer the world, albeit slowly. As late as 1997, microcredit reached only 7.6 million families, 5 million of whom were in Bangladesh alone. But in the last decade the "microfinance revolution" has come into full swing. In 2006 as many as 100 million families were reached worldwide, most of them in South Asia (Yunus 2007).

Microfinance programmes have especially benefited women, enabling them to grow their businesses, empowering them and giving them a voice in decision making (Gonzalez and Rosenberg 2006; Fisher and Sriram 2002). Most microfinance borrowers are from urban areas. They are predominantly self-employed

Box 2.1 Traditional grassroots savings and lending groups

Around the world there are many traditional forms of savings and lending groups. Here are two.

A common form is the **rotating savings and credit associations** (ROSCAs). In these, community members voluntarily come together on a regular basis; each time they meet they contribute a fixed amount of money. These contributions (or savings) are paid out immediately to one member of the group who is present at the meeting. Who gets the money is often decided through a lottery. The money enables this member to invest, buy equipment or consumer items, or pay school fees. Once a member has received a payment he or she will not take part in the coming lotteries. However, he or she will continue to participate in the meetings and pay contributions until the cycle ends. A ROSCA cycle ends once all members of the group have received a payment. At this point new members can enter the group, old members can leave, and a new cycle of contributions and payments begins. The system is particularly popular among women (Gonzalez and Rosenberg 2006; Klerk 2008).

Another common system is **accumulating savings and credit associations** (ASCAs). These go one step further than the ROSCAs. Unlike the ROSCA system, savings collected during regular meetings are not paid out immediately. They are set aside until enough money has been accumulated to provide members with larger loans or until members need to borrow money. ASCAs are more difficult to manage than ROSCAs, because the money is held as cash and records are needed to keep track of loans and repayments. This is probably the reason they are less common than ROSCAs (Klerk 2008).

One of the advantages of self-help groups is that they are self-sufficient and initiated by their members. Because members select co-members themselves, the group coherence is assured. Membership of such group also has an intrinsic value in itself as an important social resource. Nevertheless, these groups also have their limitations: their life-cycle is limited, they require good leadership (which is not always present), and their loans are generally small and short-term (Klerk 2008).

entrepreneurs: shopkeepers, street vendors, artisans, and small service-providers such as car mechanics. In rural areas, microfinance reaches clients that are mostly engaged in off-farm activities, such as food processing and trade.

So far only few microfinance institutions successfully serve farmers, though there are some notable exceptions. For example, in 2007, Grameen Bank provided a large number of loans to farmers to invest in milk cows and paddy cultivation. Also in this book, there are examples of microfinance organizations working in the countryside, for example in Ethiopia (soybean, page 164) and Rwanda (rice, page 135).

The microfinance industry offers different types of financial products (International Year of Microcredit 2005; De Klerk 2008):

- **Microcredit** means making small loans to low-income entrepreneurs so they can develop small businesses. Microcredit has helped large numbers of poor people to overcome problems due to irregular and undependable income, and to smooth their cash flows.
- **Micro-savings** are deposit services that allow people to save small amounts of money for future use, often without minimum balance requirements. The

conventional way of dealing with emergencies and expenses is to sell valuable assets such as livestock and equipment. If farm families have access to savings facilities, they can put aside part of their earnings to meet future expenses, so avoid going into debt or having to sell their assets.

- **Micro-insurance** includes life insurance for entrepreneurs and their employees and, on a smaller scale, crop insurance. It is a system by which people and businesses make payments to share risks.

Initially most initiatives were financed by donor agencies. They provided guarantees to support long-term commercial loans to microfinance institutions so these could build more capital resources and develop new financial services for small-scale businesses. In the late 1990s, many microfinance institutions transformed themselves into formal financial providers operating on commercial terms. By then it had been proven that the poor are creditworthy, so commercial banks became more and more engaged, supporting microfinance institutions with capital to increase their outreach and efficiency. Despite the growing interest from private investors, 53% of the $11.7 billion that was committed to the microfinance industry in 2008 came at below-market rates from donor agencies and multilateral banks (*The Economist* 2009).

Alongside the microfinance industry there are many traditional, informal types of savings and lending groups, which are not officially registered. Also credit unions and farmer cooperatives cooperatives remain important in providing financial services to the rural poor.

Limitations of microfinance

Microfinance has been truly revolutionary in proving that the poor are bankable, and also in allowing poor people to signal their creditworthiness. But microfinance is not a panacea. Here are some of its limitations:

- **High costs** Microfinance remains a costly service. Unlike conventional banks, microfinance organizations deal with very large numbers of small loans and savings. Handling this type of business is more expensive than dealing with a small number of large loans.
- **High interest rates**. At up to 36% a year, interest rates charged by microfinance institutions are higher than the rates charged by commercial banks (but generally much lower than those charged by moneylenders). These rates are high because of the high costs involved in microfinance operations: they need to cover the cost of the money, compensate for loan defaults and transaction costs (Kiva 2009).
- **Small amounts only**. Microfinance institutions lend small amounts of money. Experience has shown that it is better for new clients to start with small loans –generally between $35 and $800 (Yollin 2007). Such amounts can enable a borrower to make small investments in a store or to produce handicrafts, but

are generally not big enough to allow these businesses to grow to an efficient scale.

- **Short-term loans** Most microfinance programmes provide short-term loans only. They require their clients to repay relatively quickly – often within 3–4 months – and in monthly instalments. This approach is well adapted to activities where an investment pays off immediately, but is less suited to activities where the turnover of capital is slower, as in farming.
- **Little flexibility in loan conditions** Generally financial services provided by microfinance institutions do not address any special needs that borrowers might have, and they do not respond to changing conditions. Grameen Bank has responded to this problem by giving borrowers more options as to how and when they repay their loans – making it easier to pay back more during the peak business season and less during the slack time. Microfinance institutions have also responded by offering a wider range of products and services.
- **Importance of savings underestimated** Many poor people need savings more than credit facilities (De Klerk 2008). Savings help the poor by smoothing their income over time, and function as a kind of insurance. This is especially helpful when someone has an irregular income. Savings help people meet the costs of occasional events, such as funerals, and can be used for investments without having to rely on loans. Savings deposits also benefit microfinance institutions in several ways: they make them less dependent on external finance sources, help them to analyse the creditworthiness of their clients, and can be used as a guarantee for an outstanding loan, so reducing its risk.
- **Exclusion of the poorest** The emphasis within the world of microfinance has been very much on sustainability. This means that when they select clients, microfinance institutions have to choose those who are able to repay their loans. Unintentionally, that means that the poorest people cannot access financial services, and rural outreach is limited. Many microfinance institutions are trying to improve their outreach, including to rural areas, and have made tremendous efforts to improve their social performance.
- **Exclusion of others** Microfinance is not for everyone. For example, individuals often cannot get microfinance services provided to groups. That excludes people such as migrants, who may be unenthusiastic about forming groups or may lack the social linkage to do so.
- **Limited reserve of funds** Without savings, which help to generate a reserve of funds, microfinance institutions are very vulnerable and cannot respond well to periods of economic crisis. Making money more readily available would help microcredit programmes get through their initial years until they reach the breakeven level. A problem that local banks face is that they cannot lend money to microfinance institutions because the latter cannot provide collateral.
- **Professionalization** Professional staff members are needed to assess their clients' loan applications and ensure that effective loan monitoring and fi-

nancial management systems are established (De Klerk 2008). To improve information on the bankability of the lenders, some microfinance institutions have introduced credit rating and financial information systems, thereby reducing the risks of non-repayment.

- **Impact on the poor** It is difficult to measure the impact of microcredit on poverty reduction. Because microfinance institutions are largely still financed by public money, this has raised questions on the effectiveness of the small loans. Most evaluation methods fail to control for what would have happened in their absence (*The Economist* 2009).

In the countryside: Even more limitations

Perhaps the most critical limitation is that microfinance has reached relatively few people in the countryside, especially in Latin America and Africa (India is an exception, where microfinance institutions have been focusing more on rural areas). Microfinance institutions have tried to expand their services to rural areas, where poverty is concentrated and where most people are engaged in agriculture. But many barriers remain:

- **High transaction costs** Rural areas typically face high transaction costs compared to urban areas, where the density of people is higher, infrastructure is better developed and bank's branches are less expensive to maintain. The large distances to and between (potential) clients is a big challenge for microfinance institutions.
- **Lack of information for credit assessment** In rural areas information to assess a borrower's ability and willingness to repay a loan is difficult and expensive to obtain. Although group-based systems in rural areas work well in assessing a client's chracter, it is difficult to do the same for the agricultural risks involved. Products are often perishable, and it is hard to forecast price movements accurately. Key lending techniques, such as regular repayments and savings, are less suitable for agriculture, where cash flows are generally very irregular throughout the year.
- **High risks in agriculture** Another difficulty is the high risks involved in agriculture, the main source of income for the majority of the rural poor. Plus, if a drought or disease hits, or if the sale price of the produce falls, many farmers will find it difficult to repay their loans, all at the same time. Finance specialists call this "covariant risk" (De Klerk 2008: 20). It is why microfinance organizations are reluctant to extend their services to marginal and risk-prone agricultural areas.
- **Larger loans required** Large enterprises can get loans from commercial banks or equity finance from venture capital funds or the stock market. Very small entrepreneurs can get loans from microfinance institutions. But small and medium-scale businesses fall somewhere in between (KIT and IIRR 2008). These entrepreneurs have financial needs that are generally too large for microfinance, but too small for commercial banks. For example, for buying

in bulk, storing produce, or collective processing they need more financial assistance than the loans offered by microfinance institutions.

- **Inappropriate terms for loans and repayment** Most microfinance programmes provide only short-term loans. That has some benefits: problems with repayment can be identified early on. But it does not always work for borrowers in the agricultural sector, who can repay loans only after selling their crops or livestock (De Klerk 2008). So the terms used by microfinance institutions are less effective for clients with irregular cash flows, as is often the case in agriculture.
- **Distortion from government-subsidized credit programmes** Government and donor intervention in agricultural lending is one of the greatest sources of risks for agricultural lenders. For example, such interventions may offer subsidized loans (undercutting the rates offered by microfinance institutions), or may be lax on demanding repayment (undermining borrowers' willingness to repay other loans too). Politicians campaign for debt forgiveness, and governments may force lenders to reduce their interest rates. These distortions have contributed to dramatically poor repayment rates, a poor allocation of credit, and unprofitable lending.
- **Exclusion of the poorest** Excluding the poorest people hits rural areas hardest as that is where poverty is concentrated. Problems with illiteracy and age hinder access to financial services: illiterate and older people are less likely to be informed about such services or be able to understand or apply for loans.

Many microfinance institutions have recognized these limitations and are trying to overcome them, within the constraints imposed on them by national regulations. For example, they are increasingly investing in social performance monitoring to measure their impact, seeking ways to address the needs of agricultural producers and the rural poor, adding extra financial services to their microfinance support activities. Several of the cases in this book reflect these promising initiatives.

Looking at value chains

Unpredictable weather, dodgy infrastructure, volatile prices, low status, little support: despite all these problems, millions of farmers, traders, service providers and other micro-entrepreneurs still manage to deliver fresh food every day to urban consumers, export produce to distant markets, and stay in business. That reflects their resilience, creativity and huge entrepreneurial potential.

That potential becomes fully clear if we look at the value chains that link farm production to rural trading and other sectors of the economy. These chains show that farmers do not operate in isolation, but are part of a wider system. The small-scale businesses of asset-poor farmers at the beginning of the chain are intimately connected with larger businesses of traders, food processors and supermarket chains at the end.

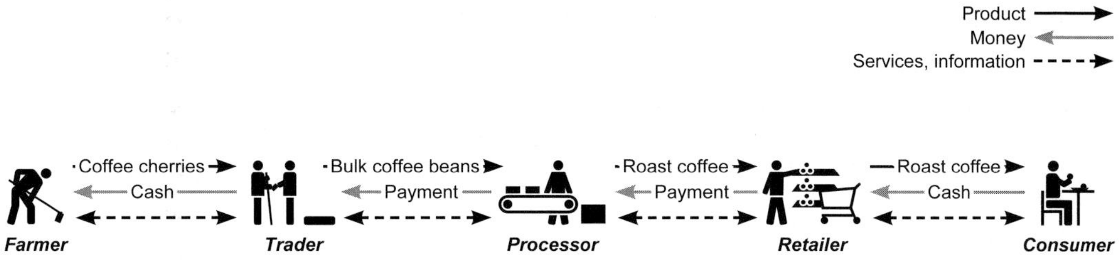

Figure 2.1 A simple value chain

A value chain refers to the entire system of production, processing and marketing of a particular product, from inception to the finished product (Figure 2.1).

A value chain consists of a series of chain actors, linked together by flows of products, finance, information and services.

Chain actors

The **chain actors** are the individuals or organizations that produce the product, or buy and sell it. In Figure 2.1, the farmer produces dried coffee cherries and sells them to the trader in exchange for cash. The trader bulks the dried cherries from several farmers, removes the pulp to produce coffee beans, and sells a lorryload of beans to the processor. The processor roasts the beans, seals them in packages, puts them in cardboard boxes, and sells them to the retailer. The retailer displays the packages on supermarket shelves and sells them to consumers.

Chain actors actually own the product at some stage in the chain. A processor who buys coffee beans and roasts them before selling them is a chain actor. A processor who roasts the beans for a trader in return for a fee, but who does not become the owner of the product, is not a chain actor, but is a **service provider** or **chain supporter** (see below).

At each stage of the chain the value of the product goes up, because the product becomes more convenient for the consumer – after all, the customer in a coffee shop in Amsterdam does not want to go all the way to a coffee farm in Togo to buy raw coffee cherries. Besides the transport costs involved, the product is not ready for use yet: the pulp has to be removed and the beans must be roasted before a customer can enjoy a cup of Togolese coffee in the coffee shop. The same is true for other crops (see also KIT, Faida MaLi and IIRR 2006).

Besides value, costs are added at each stage in the chain. For example, a farmer who dries and processes the coffee after harvesting the cherries can get a higher price – a reward for the extra work. The trader employs workers to load and transport the beans. The processor incurs the costs of roasting and blending the beans, packaging and promoting them, and distributing them to retailers. The retailer bears the costs of putting them on shelves and selling them to consumers. Costs also arise through losses that occur along the chain.

Flows of products, money, information and services

When a farmer sells a product to a trader, two things change hands: the **product** goes in one direction, and **money** goes in the other. This exchange is repeated at each stage in the chain, forming two parallel flows, of produce (the black arrows from left to right in Figure 2.1) and money (the grey arrows from right to left).

In addition, each of the actors may be prepared to invest in the chain and to support the other actors to make sure that it functions smoothly. This gives rise to additional flows of finance between the different actors in the chain. These flows may go in either direction. For example, a trader may give a loan to a farmer at the start of the season so the farmer can buy inputs such as seeds and fertilizer. Or the farmer may give the trader a loan – this is essentially what happens when the farmers get paid several months after they deliver the produce to the trader. Such financial flows may also include fees paid by a farmer to an association or cooperative that markets their produce.

In addition, the farmer and trader exchange **information**. The trader may tell the farmer how much coffee he or she wants to buy, when and where to deliver it, and what quality it should be. The trader may train the farmer on things like quality standards or new varieties. The farmer may tell the trader what the yield is likely to be and when the harvest will be ready. The two are likely to haggle over the price.

The farmer and trader also provide **services** to each other. The farmer may dry the produce, sort and grade it, put it into sacks and take it to a convenient pick-up point. The trader may provide labour for harvesting, supply sacks to hold the produce, and deal with the local government's paperwork. Similar exchanges also occur at each stage of the chain.

In Figure 2.1 and other chain diagrams in this book, the information and services are represented by the dashed arrows. (To avoid cluttering the diagrams, we normally omit such arrows unless they are explicitly mentioned in the text.)

Chain supporters

Flows of finance, information and services are not limited to the actors within a chain. Often other individuals and institutions are involved, surrounding the chain actors. We call these "**chain supporters**" (Figure 2.2).

Chain supporters may provide various **financial services** to the chain actors. These supporters include moneylenders, savings and credit groups, microfinance institutions, banks, equity funds, and so on. The financial services they provide include loans, pre-financing, shareholdings, factoring, leasing arrangements, and so on (see Box 2.2). It is not just financial institutions that provide financial services; for example, an input supplier may give a farmer a loan in the form of fertilizer, in return for repayment plus interest after harvest.

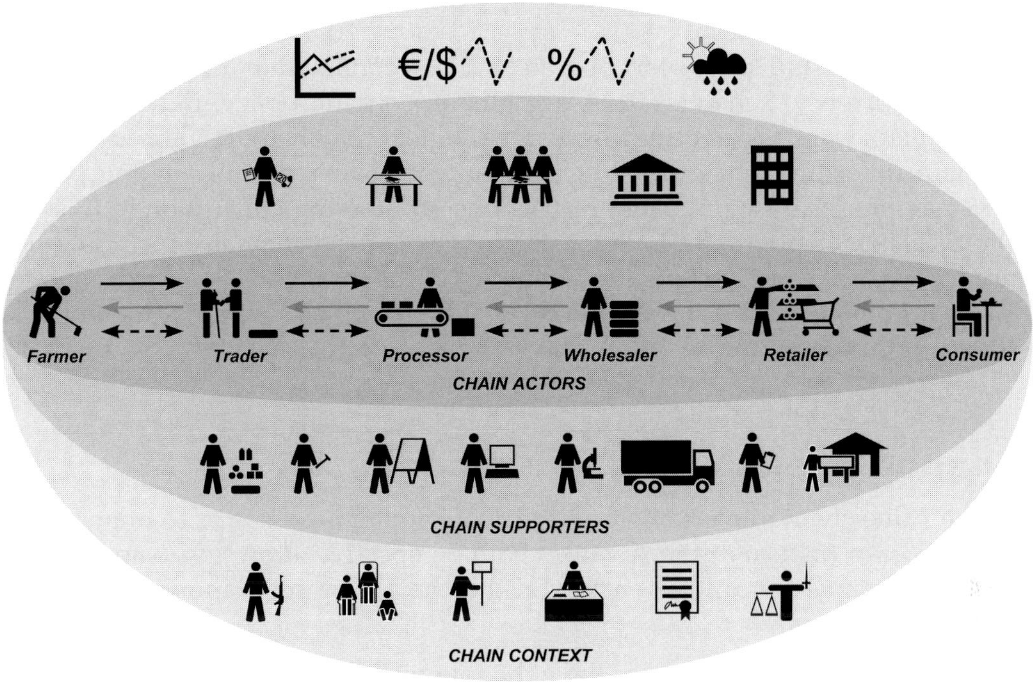

Figure 2.2 Value chain actors, supporters and context

Chain supporters may also provide a wide array of **non-financial services**: input supplies, farm labour, transport, grading, processing, storage, packaging, advertising, research, training, advice, organization, and so on. These services may be vital for the chain actors to produce the product, turn it into something that someone else wants to buy, and deliver it to the consumers.

Many of these chain supporters provide services to the chain actors for a fee. Harvest labourers want their wages at the end of a hard day's work. A transport firm will only pick up and deliver produce if the trader pays the going rate, perhaps with a down-payment in advance. Input suppliers may offer fertilizer on credit, but they want to be paid eventually. Banks expect borrowers to repay loans, and charge interest to cover their cost of doing business.

Other chain supporters do not have to be paid by the chain actors, at least not directly. They include research and extension services (paid by the government out of taxes), standards organizations (ditto), and non-government organizations (paid by donors).

Some chain supporters are paid directly by the chain actors through membership fees and the like. For example, a cooperative may cover some of its costs by charging membership fees or requiring members to put in work for free.

Chain context

The chain actors and supporters operate within a context that includes the larger economy, currency exchange rates, government economic policy, and governance, tax, regulatory and legal frameworks. This context may help the performance of the chain, for example by promoting a transparent, stable macroeconomic policy. Or it may hinder it by imposing restrictions or allowing corruption to flourish (Quirós 2006; Shepherd 2004; OECD 2006). The context may also include influence by advocacy movements (for example NGOs that work on environmental or social issues) and by social structures (for example traditional hierarchical structures in a community).

Finance needs in the value chain

For the value chain to work well, all the chain actors need access to finance. But their needs for finance differ. A value chain perspective allows us to analyse the needs for finance for different actors in the chain, and see opportunities to fill these needs.

In general, different chain actors have the following needs:

Input suppliers These provide farmers with seeds, chemicals, fertilizer and equipment, plus perhaps training or in-kind credit (such as a loan of fertilizer). Various types of input suppliers exist: big foreign-owned firms, large national companies, small-scale local retailers, farmer groups, cooperatives or public bodies – all with different financial needs. The small-scale entrepreneurs depend on small, short-term loans for working capital: they need to buy the seeds and agrochemicals and keep them in stock for their clients. Pesticides and fungicides, the more expensive products, need to be readily available for farmers when they face a problem with their crop. Farmer groups, co-ops and public bodies may depend on government funding, which may be late in arriving.

Farmers Farmers, their families and hired workers manage the crops or animals, and are involved in post-harvesting practices and marketing. Most farmers have too little money. During the production season, they often lack working capital to buy seeds and other inputs, or to hire workers to plough the land, sow, irrigate, weed and harvest the crop and to care for the animals. Especially in the months before the harvest, many farm families cannot even pay for food, household expenses or medicines. In addition, few farmers have investment capital to buy equipment such as ploughs or draught animals, or to invest in irrigation, terracing or farm buildings. So the farmers' finance needs include **loans** to pre-finance the crop, and prompt **cash payment** for their crops after harvest (or even beforehand) (KIT and IIRR 2008). They also need **credit to invest** in livestock, equipment, drying and storage facilities, and to cover the costs of labour (for example, for harvesting). If they cannot get such financial support, they will not be able to produce the quantity and quality that the buyers need, diversify their output, stay competitive, or increase their share in the final value of their products (UNCTAD 2004).

Farmers can turn to few options for money. They are seen as unbankable, and even if they do qualify for credit, the bank is a long way away, bank procedures are lengthy, and repayment systems are strict. High interest rates and transaction costs are other reasons for not knocking on the bank's door. Informal loans from business partners within the chain are often cheaper and more convenient - although they are short-term and provide small amounts, so are not enough to make investments that do not pay off immediately. Informal loans are not always convenient, for example moneylenders can lend money to farmers at excessive interest rates (perhaps 100%), putting farmers in a vicious debt-cycle.

Farmer organizations Cooperatives and farmers' associations have been one way of delivering credit to farmers, with loans often tied to farm inputs and machinery. However, like other semi-formal institutions, co-ops suffer from flawed administrative controls, lack of independent decision-making, inflexibility and high administrative costs (UNCTAD 2004). Apart from such loans, co-ops face various other financial needs. They need to cover their administrative costs. Those that market their members' produce need cash to pay their members promptly, which requires working capital; if farmers do not get paid quickly, they may sell to a private trader who pays less but who can provide fast cash. Farmer organizations that function as collection points need to invest in warehousing and transport.

Traders The traders buy produce from the farmers or co-ops and bulk it before selling it on. Their business depends crucially on making their working capital flow as quickly as possible in buying and re-selling produce. Every transaction offers an opportunity to make a profit (and, of course, carries a risk of losing money). Small rural traders have to stop buying when they run out of cash, leaving farmers stranded with their products. The traders need working capital to optimize their turnover and keep transaction costs down (UNCTAD 2004). They also need longer-term investment capital so that they can buy a vehicle, build a warehouse, or pay for equipment to weigh or grade a product. Because so much of their capital is tied up in products at any time, traders have little collateral, so find it difficult to get loans. Few financial services are designed specifically for traders.

Processors Small-scale processors may also lack the working capital they need to buy bulked products from a farmer group or trader. They often lack the money to invest in equipment, leading to losses, lowering quality, and pushing up the cost of processing. They typically need access to medium-term loans and the ability to lease equipment. Commercial banks are becoming involved in lending to such processors.

Wholesalers and exporters These sell the processed product to local and global retailers, supermarkets and corner shops, who in turn sell to consumers. Wholesalers often manage credit relations in two directions: they provide money to trusted traders so they can buy on their behalf, and they may provide products to retailers on credit, expecting to be paid after the retailer has sold the goods. In this way, wholesalers often act as a "bank" for other actors in the chain. They often need more capital than other traders in the value chain. To avoid bad debts, they need good information on the reputation and financial status of their suppliers and buyers.

Wholesalers and exporters have access to the financial services of commercial banks. These loans can be long term, or at least medium term. Exporters may have the option to provide guarantees to their suppliers (for example if they apply for a bank loan), based on an export contract. Exporters (or importers) can also participate in a joint venture, together with other chain actors. Through an equity investment they become (partial) owner of a corporation (KIT and IIRR 2008).

Financing value chains

In general, there are three types of finance for the actors in the value chain:

- **Chain liquidity** Short-term loans from suppliers or buyers within the value chain
- **Agricultural finance** Financial services from commercial banks, microfinance institutions and other financial institutions
- **Value chain finance** Financial services that are based on cooperation in the value chain.

We will discuss each type in turn.

Chain liquidity

Osmine spreads out her freshly harvested coffee cherries on the road outside her house and leaves them for several days to dry in the sun. She then scoops the shrivelled, dried cherries into sacks, heaves the sacks on her bicycle, and takes them to a local trader. She was able to get the coffee harvested on time because the trader gave her a small loan to hire some labourers. In return, Osmine agreed to sell the bulk of her output to the trader.

* * *

Such financing within a chain is common between farmers or farmer groups and traders. These credit flows are generally called **trade credit**, or **chain credit**. They consist of short-term loans to ensure a smooth flow of products, keep the chain running and maintain long-term relationships between trusted business partners. They may be given in cash or in kind. For example, a trader may provide a farmer with fertilizer at planting time, then deduct the cost from the amount he pays when collecting the harvested crop (De Klerk 2008).

Other actors may also be involved, and the credit flow can go in either direction, depending on market conditions. For example, a wholesaler may pre-finance a travelling trader; an input supplier may provide inputs on credit, and a farmer may accept a delay in payment after delivering the crop. Such types of finance

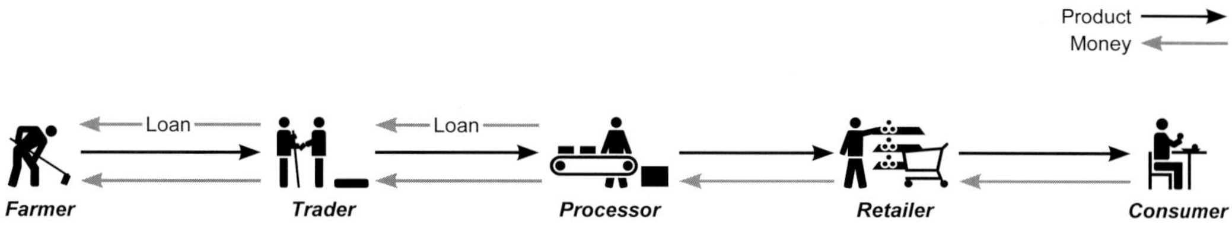

Figure 2.3 Chain liquidity

work because the actors in a chain are interdependent and share, at least to some extent, a common goal. In this book we call this type of value chain finance **chain liquidity** (Figure 2.3).

Chain liquidity has various benefits:

- **Low costs** The lenders normally do not charge interest, and the loans are informal and uncomplicated. However, this is not always the case (see below).
- **Low risk** The borrower and lender know and trust each other, and the lender knows that the borrower is likely to be able to pay the loan back. Sometimes, the main objective of the loan is not to make money but to maintain relationships with chain partners.
- **Tailor-made** The loan terms are based on the individual business deal. The amount and terms of repayment can be varied depending on the needs of the two parties.
- **Improved chain efficiency** This is a win–win situation. Both the borrower and lender have lower transaction costs and fewer risks as a result of the loan: the borrower can use the money to produce more and gets a guaranteed buyer, while the lender gets a more reliable supply. The loan may reduce the time and costs of transport and communication, as well as losses caused by an actor not being able to sell perishable stock (KIT and IIRR 2008).

Chain liquidity is easy and convenient, and provides many benefits to buyers and sellers. But it does not solve all their financial needs. Some limitations of chain liquidity are:

- **Relies on trust** Chain liquidity works only in long-standing relationships and is built on trust.
- **Short-term, small amounts** Chain liquidity only concerns short-term loans and relatively small amounts. The duration of the credit is no longer than it takes to buy and sell the product. The loans cannot be used for investments.
- **Danger of dependency** Chain liquidity lacks transparency, may have high "embedded" costs, and there is a risk of dependency and exploitation. Many cotton farmers in India have become trapped in a vicious cycle of debt by buying bad seeds and costly pesticides that yielded little due to drought. Unable to see a way out, more than 25,000 farmers have committed suicide

since 1997, many by drinking the chemical that was supposed to make their crops grow.

Agricultural finance

Johnson, the trader who buys Osmine's coffee, needed working capital to buy dried coffee cherries from many farmers, and to hire a lorry to take the coffee to a processing plant. So he has arranged a loan from a microfinance institution. At the end of the harvest season, he will pay back the loan, plus interest.

* * *

This is an example of an outside agent (the microfinance institution) offering specialized financial services to an actor in the value chain (the trader). This is what we call **agricultural finance** or **external finance**. Here, microfinance institutions, banks and other financial agents become chain supporters in one-to-one relationships with a chain actor (Figure 2.4). Commercial banks have started to use the techniques of microfinance institutions to lend to rural entrepreneurs.

Agricultural finance may take various forms and target different actors in the chain. It includes loans, deposits and insurance. These specialized financial services are longer-term than services falling under chain liquidity. Also, they involve larger amounts of money, they are more transparent, and the risks of exploitation are considerably less. But agricultural finance also has its limitations. The section on microfinance at the beginning of this chapter discussed some of the difficulties in providing financial services to the agricultural sector: high transaction costs, a lack of information on borrowers' creditworthiness, high risks, and the bad past performance of rural credit organizations.

Other limitations of agricultural finance are:

- **Lack of flexibility** Rural finance is embedded in underdeveloped financial markets, with cumbersome procedures, strict requirements for lending, high interest rates, and high transaction costs. Microfinance institutions generally have too little capital to provide investment capital to small and medium enterprises. Commercial banks generally lack incentives to do so. In many African countries, banks are owned wholly or partly by the government or large business holdings, and tend to allocate credit to favoured sectors or affiliated companies.
- **Bankability of small businesses** Rural businesses are often informal, lack written records or formal accounting, and have little capital which can be used as collateral. Financial institutions are reluctant to lend to them. Requirements for deposits, collateral or balance sheets often act as insurmountable barriers to small- and medium-scale farmers and traders. In Cameroon, for example, the minimum deposit needed to open a checking account with a commercial bank is over $700 (Yago et al. 2007).
- **Inadequacy of formal finance** Farmers and traders do not always want formal finance. They often face immediate needs for cash that are incompatible with lengthy bank procedures and with strict repayment systems. High

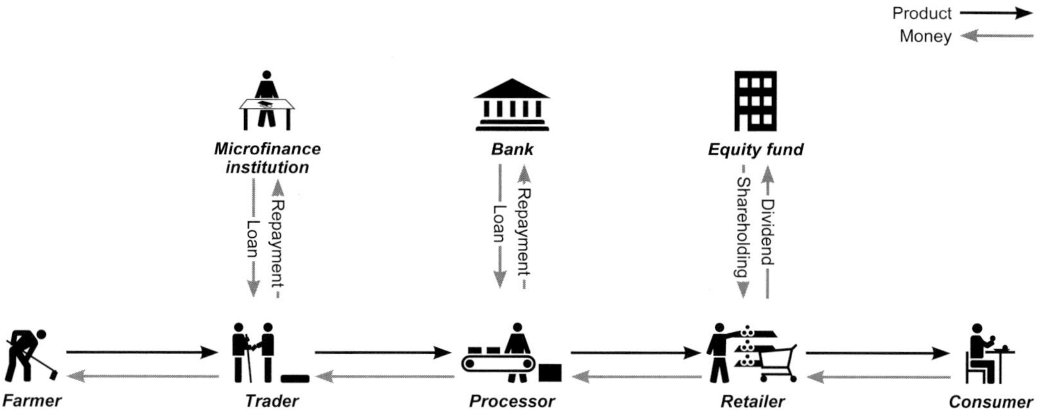

Figure 2.4 Agricultural finance

interest rates and the costs of applying for the loan are other reasons for not walking into the bank. Informal loans from business partners are often cheaper and more convenient.

- **Lack of access** Rural entrepreneurs still have very limited access to this type of finance Banks and microfinance institutions have few branches in rural areas or expertise in the agricultural sector. Few offer financial services that are adapted to the needs of farmers and traders.

Value chain finance

Johnson's business was growing, and he needed more working capital to buy more coffee from Osmine and her neighbours. He went to a local bank to arrange a loan. He had no hard collateral to offer, so the bank agreed instead to rely on soft collateral: Johnson's relationships with the farmers and with the processor. It is confident that Johnson will be able to repay the loan because he has bought from the same farmers for several years, and is a regular supplier to the processing plant. As a condition of the loan, Johnson, the processor and the bank sign an agreement: the processor will pay Johnson not directly, but through the bank. The bank deducts interest and part of the loan principal before paying the remainder of the money into Johnson's account. To make sure things work smoothly, Johnson tells the bank how much capital he needs, and the processor tells it how much coffee he has delivered.

* * *

Value chain finance is when one (or more) financial institutions link into the value chain, offering financial services which build on the relationships in the chain. The seller, the buyer and the financial agent work together, using the business relations in the value chain as a carrier to provide financial services. In the example above, Johnson gets working capital from the bank, which is repaid by the processor. This form of finance provides benefits to all parties:

Box 2.2 Some examples of value chain finance

Warehouse receipts In this system, farmers take their produce to a warehouse and get a receipt in return. They can use this receipt as collateral if they want to apply for a loan, so do not have to wait for payment. This is a useful arrangement for co-ops that want to store their products until prices rise, or if farmers have to wait for payment from buyers.

Repo finance Repurchase agreements ("repos") are a form of commodity finance. The bank actually buys the product from the seller (e.g., a co-op), and at the same time signs a contract to sell it back to the co-op at a certain point in the future. The contract specifies a price that reflects the costs that the bank incurs.

Private equity A bank or other investor may buy shares in a company, so giving it capital it can use to invest.

Leasing This is an alternative to long-term loans to buy equipment, which many financial institutions think are too risky. The leasing company provides the farmer (or other borrower) with equipment for a few years on a contract basis, and the farmer pays off the lease in instalments. At the end of the lease period, the leasing company either repossesses the equipment or offers to sell it to the farmer. Leasing is less risky than a loan because the equipment remains the property of the owner, who can withdraw it easily if the farmer defaults on payments. With a loan, by contrast, it may be difficult to take possession of the collateral offered to guarantee a loan because legal constraints and weak judicial systems (Klerk 2008; KIT and IIRR 2008).

Factoring A farmer delivers the produce to the buyer and writes an invoice for the amount delivered. Instead of asking the buyer to pay, the farmer sells the invoice to a third party, a factoring house. The factoring house pays the farmer immediately (minus a fee), then submits the invoice to the buyer for payment.

- Johnson has better access to capital to cover his financial needs. The fact that it is provided by a bank gives him more independence than if he got a loan from the processor.
- The processor does not need to engage in financial services and deal with the problems of disbursement and repayment. Rather, it can focus on its core business.
- The risks of the bank in financing Johnson are lower because the loan is securitized by the agreement between the three parties, and the loan will be repaid by the processor.

Box 2.2 gives some more examples of value chain finance, and Figure 3.1 (on page 28) shows it in diagrammatic form.

Trust is key in value chain finance. Trust is related to the duration of relationships and the degree of openness with which the chain partners exchange information. The more trust between the business partners in the value chain, the better are the conditions for good business performance. At the same time, when the chain partners share information on a frequent basis, this will also contribute to the bank's understanding of how the value chain works. In-depth knowledge of the value chain makes risks better manageable, so the bank will be more willing to engage with the value chain and take the risk of lending to asset-poor farmers, traders and other rural businesses.

Value chain finance is not a goal in itself, and only seldom is it the ultimate solution for a problem in a chain. Indeed, often the most severe constraints in value chains are not financial: they include a lack of technical knowledge or the absence of an attractive market. But value chain finance can help to set up a value chain, smooth out bumps in it, or upscale the operations in the value chain, so increasing the chain's competitiveness. It complements (rather than replaces) existing credit flows between actors in a chain, and it empowers the chain actors by making new sources and forms of finance available to them.

In the next chapter we will look in closer detail at value chain finance, and present a framework for how we analyse value chain finance in this book.

3

A value chain finance framework

IN THE PREVIOUS CHAPTER we saw that value chain finance results when financial agents link up with two or more value chain actors. Value chain finance thus combines two worlds – the world of value chain actors with the world of specialized financial service providers. The first is a world of soil management, crops, irrigation, harvesting, transport, trading, consumer marketing, etc. The latter is a world of balance sheets, risk management, credit rating, guarantees, collateral, and so on. The more these two worlds are connected, the more they benefit from each other:

- **Smallholder farmers, traders, processors and other chain actors** can improve their access to financial services. Previously they might be excluded from bank loans due to a lack of collateral, a perception of high risk, or other barriers. But now they can obtain external finance based on the fact that they have a small but viable business that forms part of a wider and more stable system of value creation – the value chain.
- **Financial institutions** can develop whole new markets for their services. Smallholder farmers, traders, processors and agribusinesses become bankable clients. In many developing countries, agriculture is the backbone of the economy. Therefore, the ability to serve this sector greatly increases the scope of action for financial institutions.

In this section, we take a closer look at how value chain finance works. We will present a framework to understand and analyse value chain finance. This framework will be used in the case studies that follow in the other chapters.

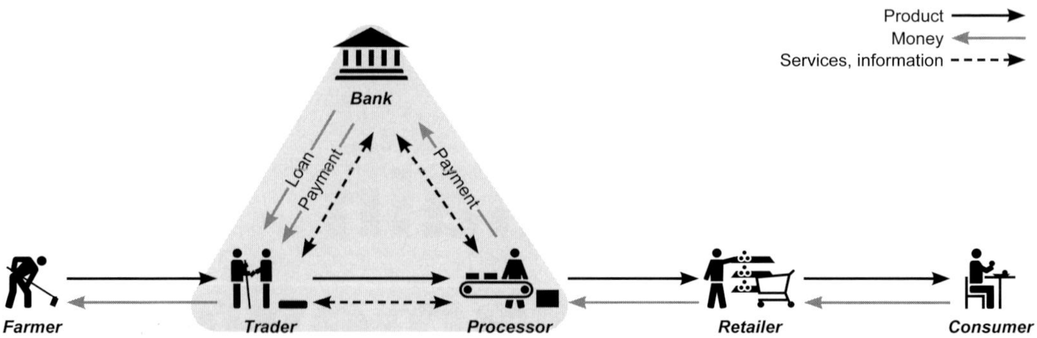

Figure 3.1 Value chain finance

The triangle of value chain finance

In value chain finance, a financial institution engages with the actors in the chain. This creates a triangle of cooperation (Figure 3.1). The triangle is between the seller, the buyer, and the financial institution.

Together they make an agreement which covers four essential aspects:

- The **product** that is produced and sold
- The **finance** needed to produce and deliver the product
- The way the parties communicate and exchange **information**
- The way **risks** are managed.

Below we explain the four aspects in more detail, using another example of a farmer selling coffee to a processor. Keep in mind that the buyer and seller could also be other chain actors, like traders, wholesalers or retailers.

Product flow Before the season starts, the farmer agrees to produce and deliver coffee to the processor. Preferably this is recorded in a contract. The batch of coffee the farmer will deliver at the end of the season is specified as precisely as possible to determine its market value. Also, the parties estimate the costs needed to produce, harvest and deliver the coffee. In this way, the parties calculate both the loan size that the farmer needs for working capital, as well as the collateral value of the coffee. Normally a loan would not be more than 70% of the market value of the product, in order to prevent repayment problems in case the harvest is disappointing.

Financial flow After the purpose and the size of the loan are determined, the parties agree on other important aspects of the financial product. These include:

- **The interest rate** of the loan.
- **The timing and form of disbursement** The loan may be given at once up-front, or in small parts during the season when the farmer needs the money.

It may be paid directly to the farmer by the financial institution, or through the processor. It may be paid in cash, or in kind, or in some combination of the two.

- **The timing and form of repayment** The repayment can be done by the processor by deducting the loan from the value of the coffee the farmer delivers. Or the money may be repaid directly by the farmer;
- **The liability for the loan** The farmer may be solely responsible for repaying the loan, or it may be a shared responsibility between the farmer and the processor.

Information flow The farmer and the processor also agree on the information that they will share with the financial institution to guarantee the loan is managed well. There are two important periods of information sharing:

- **Before the loan is approved** Before the financial institution can release the funds, it needs a good insight into the operations of the farmer and his or her business relation with the processor. This is called **due diligence**. Usually the financial agency will ask for information like:
  - A signed contract between the farmer and the processor
  - Information on previous years' production
  - A detailed plan on how the loan will be used
  - A membership statement of a cooperative or other types of farmer groups
  - The farmer's identity document
  - Report of a technical audit of the farm.
- **During the loan period** After the loan is released, the financial institution will appreciate frequent information about how the business is going. This is called **monitoring**. Usually the financial institution will require a monthly report about the status of the crop, the expenditure so far, and the outlook for the rest of the season. Preferably the report is based on technical visits from the processor's field staff.

Risk management Finally, the parties agree on how the risks of the loan are being managed. This is called **securitization** of the loan. In traditional finance, the bank usually requires hard collateral in the form of fixed assets (such as land and machinery) for a total value of up to two times the size of the loan. The purpose of value chain finance is to reduce hard collateral requirements so that farmers and other chain actors can more easily access financial services. Usually in value chain finance we see the following forms of collateral:

- **The contract between buyer and seller** As mentioned earlier, normally a loan would not exceed 70% of the value of the contract.
- **The product itself** The financial institution may put a claim on the product while it is being stored or shipped.

- **Solidarity guarantees and peer pressure** The financial institution may require that peers, neighbours or family of the farmer become co-guarantors for the loan.

Hence, the triangle of value chain finance is an elaborate agreement between the buyer, the seller and the financial institution. Their agreement covers the product, the financial service, the sharing of information, and the management of risks. Each case study will closely examine how these details are arranged in the particular finance deal at stake.

We use the icons shown in Table 3.1.

Table 3.1 Icons used to depict financial products

Product and financial flows	
Purpose	What is the loan (or other financial product) used for?
Amount	What is the amount of the loan? Total of all loans, average loan per borrower, or range.
Period	What is the average period of the loan? When must it be fully repaid?
Disbursement	How is the loan disbursed? In instalments, or all at once? To whom?
Repayment	How must the loan be repaid? In instalments, or all at once? By whom?
Interest rate	What is the interest rate (per year or other period)?
Transaction costs	What other costs are incurred?
Risk management	
Securitization	How is the loan securitized? How can the lender be sure of getting the loan repaid?
Liability	Who has to repay the loan?
Information flows	
Information required to apply	What information does the lender require before the loan is disbursed?
Information required during season	What additional information does the lender require during the loan period?
Time lag between application and payment	How long does it take between an application and the disbursement of the loan?

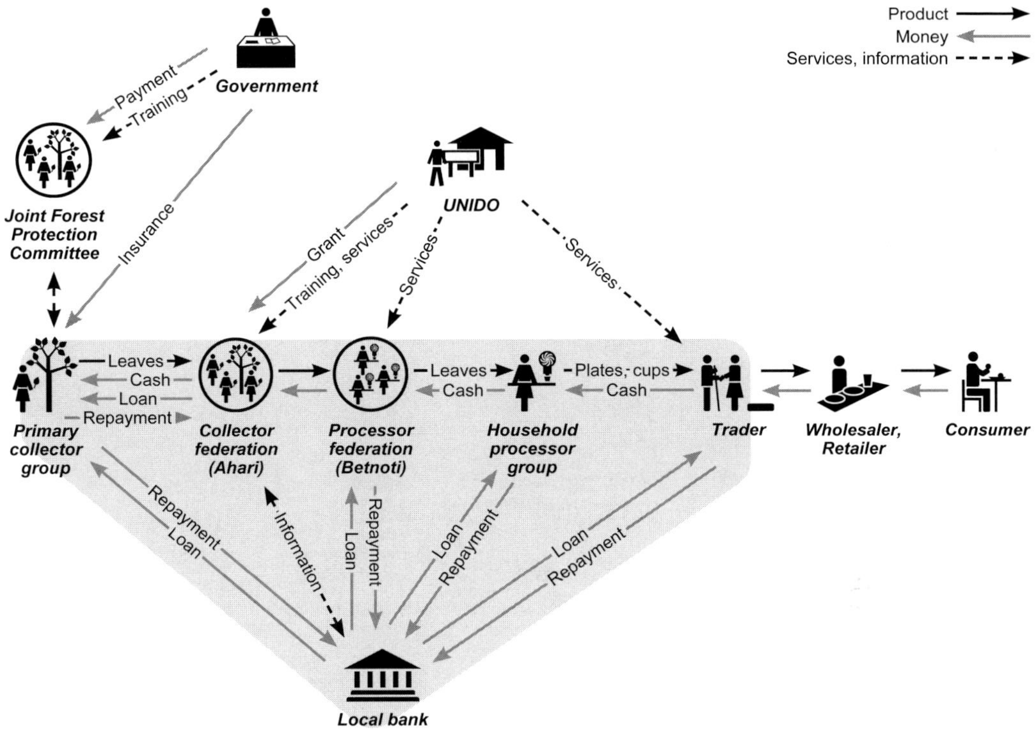

Figure 3.2 *A single financial institution finances several stages in the value chain* (from the sal leaf India case, page 116)

The chain of value chain finance

So far, we discussed value chain finance as a triangle of cooperation between the financial agency, the buyer and the seller. But in practice we often see that multiple actors from the value chain are involved. Not only the farmer, but also the trader needs finance. And not only the trader, but also the processor, the wholesaler and the retailer. For the value chain to work well, there need to be proper financial services at all stages of the value chain. This we see, for example, in the case on sal leaf in India (beginning on page 110). The local bank provides loans to:

- The women who collect the sal leaves in the forest
- The processors who make the leaves into plates
- The traders who sell the plates to the wholesalers and retailers.

This is shown in Figure 3.2. The triangle has been expanded to cover five stages in the value chain.

But it can get even more complex. Often we see that various financial institutions are involved at the same time. This is because each financial institution has a specialization. Credit cooperatives are specialized in working with smallholder farmers. Microfinance institutions are strong in working with small-scale traders. Banks are good at financing larger companies. So we often see that multiple financial institutions offer financial services to the value chain actors.

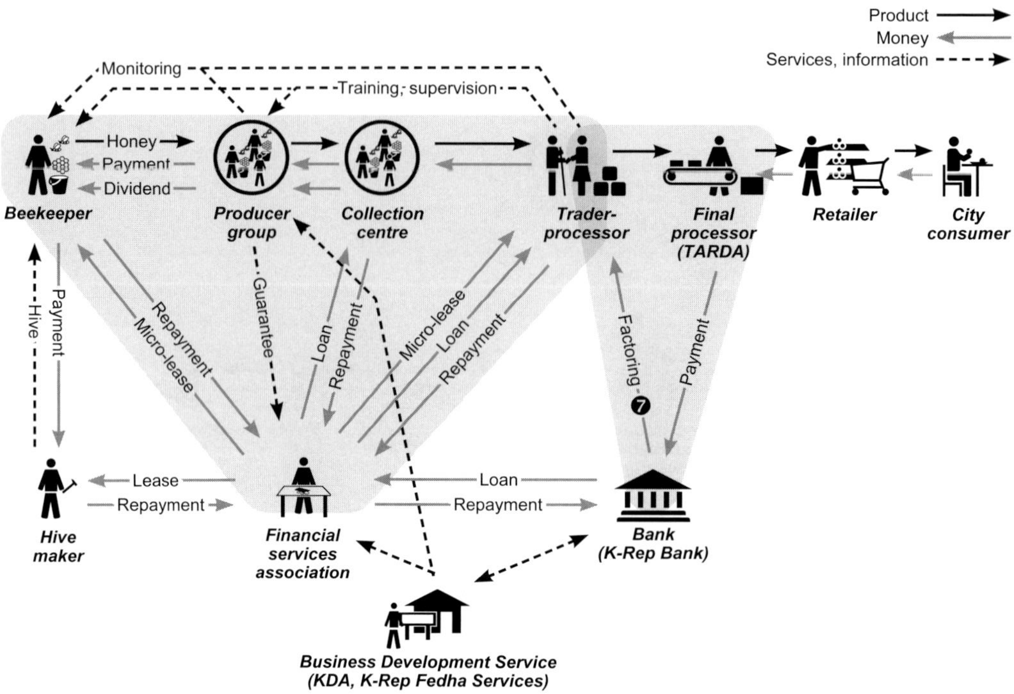

Figure 3.3 A complex value chain finance model with two financial providers
(from the honey Kenya case, page 154)

For example in the case of honey in Kenya (beginning on page 149), there are two financial service providers. There is a specialized financial service association that provides leasing and small loans to the beekeepers, hive makers, and honey collection centres. At the same time, a bank provides loans to the traders and processors (Figure 3.3). The triangles have been expanded and multiplied to cover five stages in the honey value chain.

When more actors are involved, the triangle grows to become a **chain of value chain finance**. This is because the various stages in the value chain needs various forms of financial services at the same time. Each financial services requires specific capacities, therefore also various financial agencies might be involved at the same time. Figure 3.4 shows how this may work.

In this book we will see chains of value chain finance in six case studies: milk India (page 68), sal leaf India (page 110), rice Rwanda (page 135), honey Kenya (page 149), soybean Ethiopia (page 164), and fish India (page 200). This is no coincidence. Apparently an integrated approach in value chain finance works better – you cannot strengthen the value chain unless all links work well.

Reorganization of the value chain

Value chain finance is not simply a matter of pumping money into the value chain. On the contrary, when financial institutions engage with the value chain,

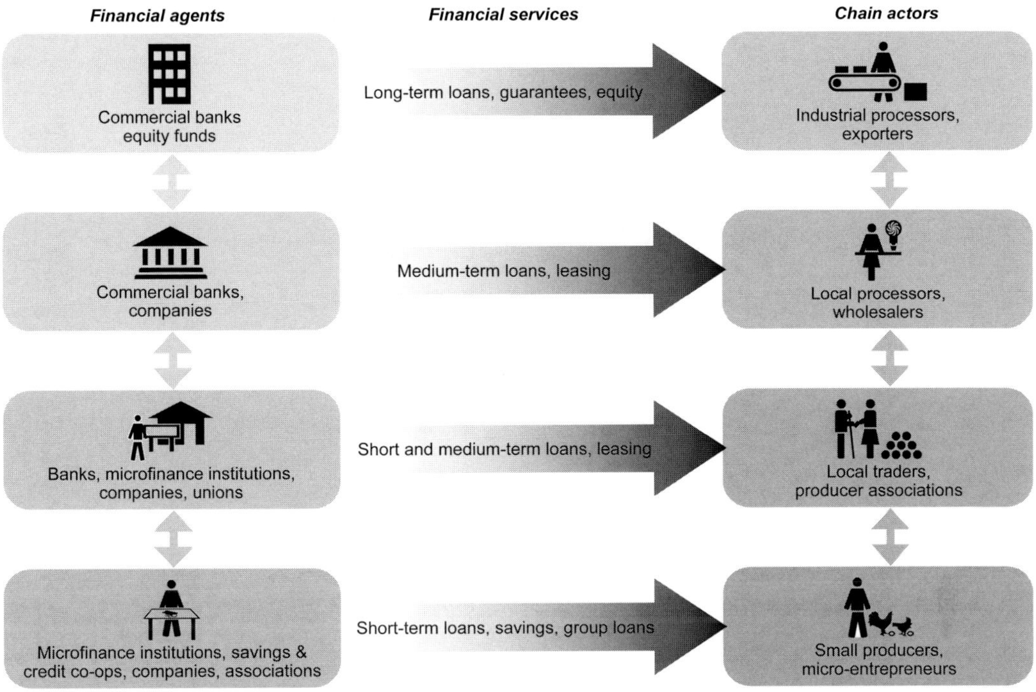

Figure 3.4 The chain of value chain finance
(based on original by Terrafina Microfinance)

they do so to make the chain more efficient and competitive. Value chain finance is a deliberate intervention that restructures and reorganizes the value chain. It triggers significant changes in the flows of money and information in the chain, in the services provided to the chain, and sometimes even in the product flow itself.

Take, for example, the case of honey in Kenya (page 149). Figure 3.5 shows how it was organized. Before the intervention, beekeeping was not a profitable business. Few people who owned hives also kept bees. These beekeepers obtained traditional hives from artisan hive makers. In exchange for these hives they would give the hive makers honey or some other in-kind payment. The beekeepers would sell the small quantities of low-quality honey they produced to local traders or directly to their neighbours. The local traders, often without cash, would borrow money from moneylenders against high interest rates to pay the beekeepers. Home brewers and herbalist would buy honey from the trader for further processing into alcoholic drinks and medicine. These products were sold in the capital of Kenya, Nairobi.

In this chain many opportunities were underused. For example, the potential for honey production was much higher than in reality. The techniques used by the beekeepers were inadequate. More efficient organization could add enormous value to the whole chain and to its individual actors. But this reorganization required support. When the bank intervened, it stimulated the rebuilding of the entire value chain. The beekeepers were organized in groups and trained how to

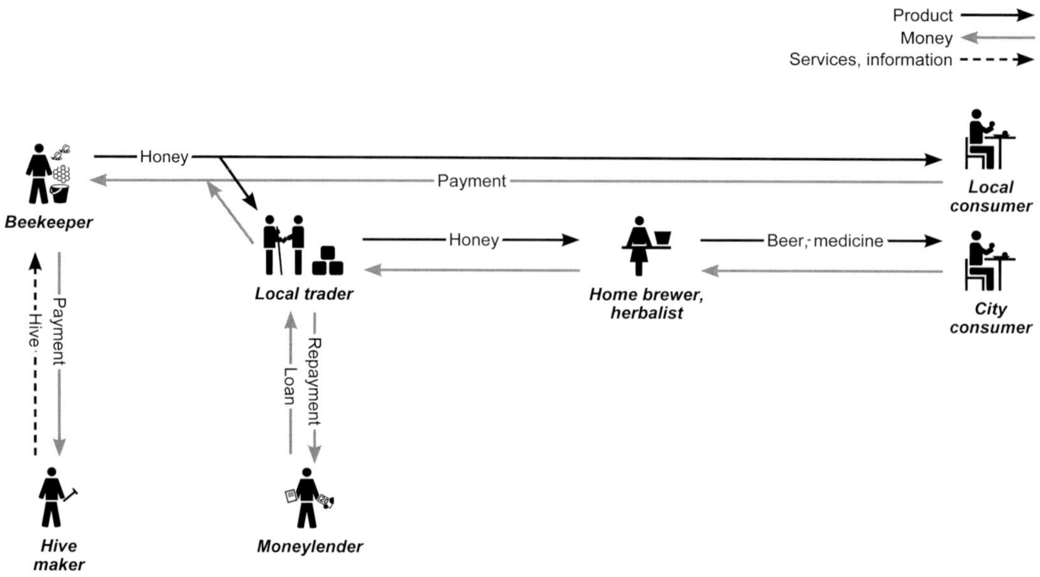

Figure 3.5 Honey value chain before intervention
(from the honey Kenya case, page 149)

improve the quality of their production method. Traders and hive makers were also trained, for example on financial management. Financial services associations (a type of village co-op) were introduced, providing a variety of financial products to different actors in the beginning of the chain. The bank itself provided financial services to actors higher up in the chain. The chain was reorganized as in Figure 3.3.

In conclusion, value chain finance is more than pumping money into the chain. Value chain finance is a deliberate intervention that reorganizes the value chain to make it more effective and efficient. Each case study captures this by presenting diagrams of the past and present situation of the value chain.

Risk management

The first part of this chapter explained that financial institutions are reluctant to provide loans to rural entrepreneurs because they perceive high risks in agricultural production and trading. The most important sources of risks in the agricultural sector are:

Production risks A farmer's crop may fail due to drought, excessive rain, pests, diseases, fire, and many other factors. A trader's merchandise may be lost due to expiration, theft, leakages in the warehouse, and many more. A processor's output may be worthless due to contamination, human errors, etc. All these are production risks.

Price risks Agricultural markets are extremely volatile. Prices go up and down, and may vary 100% or more during the same season. A trader may go to a remote area to purchase a load of tomatoes, and arrive at the market 3 days later to find several other lorryloads of tomatoes have arrived on the same day, and that prices have fallen. That trader will lose a lot of money. This is a price risk. Price risks rise when there is little market information, or when markets in one place are not connected to the markets in other places.

Market risks Closely related to price risks are the risks of market demand. When supply is larger than the demand, prices will drop, and some farmers will not be able to sell any of their produce.

Default risks Default is when somebody fails to honour an agreement. This may be a farmer who promised to deliver a product but then decides to sell it to a third party who offers a better price. Or it may be a trader who has received the product on credit but then fails to pay for it. Or it could be a processor who took a loan to invest in new machinery, but then goes bankrupt and fails to repay the debt. Obviously the risk of default is a key consideration in issuing loans.

Currency risks Farmers and processors often use inputs and technology from abroad. And many agricultural products are traded internationally. In such situations, currency risks may arise. When the exchange rate of the local currency drops against the euro or dollar (or whatever foreign currency the inputs are priced in), the farmer or processor may be faced with higher costs. If the local currency strengthens against the foreign currency, they may receive less for the produce they sell.

Other risks The political and legal environment is often a source of uncertainty. When social unrest breaks out, companies may be affected. Or governments may intervene in the market for political reasons.

Value chain finance is a way to manage and mitigate these risks on the basis of strong collaboration in the value chain. When farmers, traders, processors, retailers and other chain actors work together, they can jointly reduce the risks that each of them faces in their businesses. When the buyer and the seller make a sales contract, then the market risks are almost absent. Production risks may be lower because the buyer may give technical assistance to the farmer. The risk that farmers do not pay back the loan is lower, when repayment is done immediately when the products are delivered to the buyer.

So value chain finance has the benefit of reducing risks by promoting better coordination between the buyer, seller and financer. In financial jargon, value chain finance reduces **credit risks** into **performance risks**. The risk of a farmer being unwilling to pay off a loan (a credit risk) is much higher than that of the farmer being unable to produce a certain volume of produce (a performance risk). The risk of a trader not paying off a loan (credit risk) is much higher than that of the trader stopping his trading activities (performance risk).

As the risks are reduced, it becomes more attractive for the financial agent to provide loans to the actors in the value chain. Also, as the risks are lower, the interest rates on the loans can be reduced, which is to the benefit of the actors in the value chain.

Each case study in this book presents an analysis of the main risks in the value chain, and how these risks are mitigated through value chain cooperation. Table 3.2 gives an example how this is done in the case of organic cotton in Tanzania (see page 103).

Benefits

The premise of this book is that value chain finance provides for a triple-win situation:

Table 3.2 Mitigating risks for organic cotton in Tanzania

Type of risk	Source of risk	Risk level	Mitigation strategy
Production	Drought	●●○	Accurate yield forecasts, based on visits to farmers every 2 weeks
	Contamination of chemicals ("drift")	●○○	Grow hedges or leave gaps near non-organic cotton fields Compliance checked by inspectors
Price	Price fluctuations	●●○	Purchase guarantee from Remei with continuous price adjustments based on "open-book budgeting": Remei looks at BioRe's production costs and adds a small profit margin to calculate sales price
Market	Lack of market demand	●●○	BioRe has purchase guarantee from Remei BioRe commits to buying only 80% of actual yield of farmer
	Market may not pay premium for cotton not yet fully certified as organic	●○○	BioRe will continue to buy such cotton and pay the input subsidy, but will stop paying the organic premium
Default	Side-selling to competing traders	●○○	Price premium Close relationship with farmers based on transparency Short storage period (weekly) Budgeting accounts for 9% default
	Non-compliance with organic regulations	●○○	Visits to farmers every 2 weeks Internal and external inspection Continuous training
Currency	Exchange rate US$–TSh	●●●	Loan and payments are in dollars, so currency risk reduced to 2 months that cotton is in BioRe's hands

Risk levels: ●●● High ●●○ Medium ●○○ Low

- **Value chain actors** improve their access to financial services and can therefore improve their businesses.
- **Financial institutions** can develop new markets for their services.
- **Economic development and poverty reduction** are attained as small-scale entrepreneurs get better conditions to flourish their businesses.

To capture these benefits, the case studies analyse the benefits of value chain finance at various levels.

Small-scale farmers, farmer groups and traders gain better access to finance. Value chain finance breaks with the need to have hard guarantees. The business relations in the value chain replace the need for hard collateral. When a buyer with a reputation as a reliable purchaser is willing to vouch for its producers, even small producers become more attractive clients to financial institutions (USAID 2005). There are fewer obstacles to credit provision, the terms and services are better, and the loans reflect the cash flow pattern of the business activities (UNCTAD 2004).

Access to financial services enables farmers to invest in better equipment and input, hire workers, improve the quality of the product or diversify their income by getting involved in processing or by producing other types of products. In this way value chain finance can contribute to higher and more stable incomes. Besides loans, other types of financial services (such as micro-insurance) are being offered to farmers. These services help farmers in future planning of their business.

Value chain finance may entail more than financial services; for example, it can also involve training and the organization of farmers into groups. Training often contributes to the production of better quality items, greater efficiency and higher volumes of the product, resulting in higher prices and incomes for farmers. The organization of farmers into groups can serve different goals. It can contribute to better negotiation positions, help reach price agreements, and create access to group loans. Farmers may also benefit in less tangible ways, for example through increased levels of trust and dignity.

Small traders share some of the benefits that go to the farmers. Higher quality, higher volumes and better continuity in supply enable traders to expand their business. Direct funding to traders can support them to pay their suppliers at the moment of delivery. For many farmers immediate payment is crucial and an important reason to sell to a specific trader. It thus cements loyalty between farmers and traders.

For **processors and other buyers** the main benefits of value chain finance are to secure a supply of better-quality raw materials. Value chain finance enables the processor to invest in its suppliers and to gain their loyalty. Providing financial services directly to the processor enables it to invest in processing equipment.

Consumers benefit by gaining access to higher-quality products, a more regular supply and a greater choice of products.

For the **value chain as a whole,** financial services help to streamline and smooth relations in the chain. Trust among different actors can help revive the chain. The chain itself can upgrade because it carries a greater volume or a higher-quality product, which fetches a higher price, which will be to the benefit for all chain actors.

Because specialized financial institutions are involved, actors in the chain may have access to larger loan sums and a greater range of services, including savings, leasing, investment loans and insurance (USAID 2005).

For the **financial institution**, benefits include an increase in the number of clients and the size of a loan portfolio, a gain in the trustworthiness of clients, and improvements in the quality of financial services the institution can offer. Value chain finance mitigates risks for financial institutions and reduces their transaction costs. For example, the costs of credit screening, monitoring and enforcement are lower because the chain actors do much of this work themselves. That cuts the financial institution's costs and lets it provide loans at lower interest rates.

The product flow in the value chain is used as a carrier to provide financial services. Market risks (the risk that no buyer will be found) are almost absent, because the buyer is involved in the deal. Default risks (the risk that farmers do not pay back the loan) are lower, because repayment is done right at delivery of the farm products. Production risks (the risk that the farmer fails to produce the product) remain, but may be lower because the buyer may give technical assistance to the farmer.

Wrapping up

This chapter has presented the framework that will be used to analyse the experiences in the case studies. So much for the theory: let us now look at the experiences of people who have made value chain finance work in the real world.

4

Crafting new chains

THIS CHAPTER FOCUSES ON the use of value chain finance in building new value chains from scratch. It contains four cases:

- **Turning potatoes into crisps in Peru** shows how an NGO has helped a group of potato growers negotiate a deal to supply potatoes to a large maker of snack foods, and helped them obtain value chain financing for their production. The NGO has guided the farmers through the first few years of this business relationship, and now the farmers are able to manage nearly all aspects of the business themselves.
- **Milk helps revive a microfinance institution in Bolivia** tells how an NGO seeking to help milk producers in the Altiplano linked them with Sartawi Foundation, a struggling microfinance institution, to supply dairy companies with raw milk. The collaboration has worked: the milk producers have upped their production and now earn ten times more than before; Sartawi is now a vibrant business; and the dairy companies have reliable suppliers of milk.
- **Creating a new dairy value chain in India** is also about milk. It shows how Reliance, one of India's biggest companies, along with BASIX, a private company which promotes livelihoods for the rural poor, have set up a value chain that buys tiny amounts of milk from smallholder farmers. This case is remarkable because the decision to buy from such small-scale producers was a hard-headed commercial one, not based on humanitarian considerations.
- **Fighting elephants by growing chilli in Kenya** describes the efforts of small-scale farmers to grow crops in an area prone to wildlife incursions. The answer was chillies, along with marketing agreements with buyers, which in turn acted as soft collateral for loans from a local bank.

Turning potatoes into crisps in Peru

PICK UP A PACKET of potato crisps in a Peruvian supermarket, and chances are that it will have a "Lays" logo on it. "Lays" is a brand of Snacks America Latina Perú SRL, part of Pepsico, a big American multinational. Snacks is the market leader: it supplies 85% of the Peruvian market for crisps.

To make its crisps, Snacks uses two particular varieties of potato, "Capiro" and "Canchan." These varieties are pale yellow and high in dry matter: good for making crisps. They are normally grown high in the Andes, where the production season lasts from March to July. Potatoes are in short supply during the rest of the year. Canchan must be processed immediately, but Capiro can be kept for up to 4 months before they are sliced into crisps. Despite this, Snacks still found it hard to keep supermarkets stocked with crisps all year round.

So the firm needed an alternative source of potatoes: one that bridged the seasonal gap. This alternative was the coastal lowlands. It almost never rains here, but with irrigation it is possible to grow two or three crops a year.

Snacks may be a big firm, but it can't do everything. It is good at slicing and frying potatoes, not at training and organizing farmers to grow the crop. It was willing to pay extra for a reliable supply, and to agree on prices beforehand. But it would be too cumbersome to sign contracts with lots of individual farmers. The answer was to team up with Fovida, an NGO that promotes the development of small-scale farmers. Snacks had bought produce from Fovida-supported farmers in the mountains, so knew how the NGO worked. Fovida agreed to help develop production of Capiro potatoes in the coastal lowlands to help fill the seasonal gap.

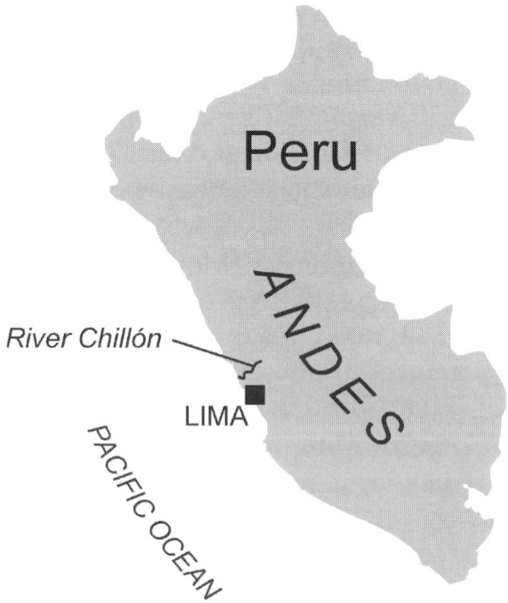

Figure 4.1 Location of the River Chillón in Peru

Developing a value chain: 2001–7

Fovida suggested working with farmers in the valley of the Chillón river, near Lima, the country's capital (Figure 4.1). The farmers were poor, unorganized, small-scale producers who were growing vegetables for sale in the local market. They also grew potatoes, though not the varieties required by Snacks. But Fovida was confident that it could help them develop a production system that would fulfil Snacks' stringent quality and quantity requirements.

Fovida helped the farmers organize into a producers' group, called the Asociación de Productores de la Zona Media del Río Chillón (Producers' Association of the Middle Chillón River). This organization has 34 members, each with between 2 and 13 ha of land, or an average of 5 ha.

At first, Snacks did not want to sign an agreement directly with the Asociación: it would be too difficult for the firm to manage such an arrangement with an unproven group of suppliers. It preferred to sign a contract with Fovida instead, making the NGO responsible for supplying the potatoes to the Snacks factory. This contract specified quantities, prices, delivery times, and so on. Fovida also was obliged to provide post-harvest services such as selection, loading and transport of the potatoes, and handle all the documentation. It also had to ensure that

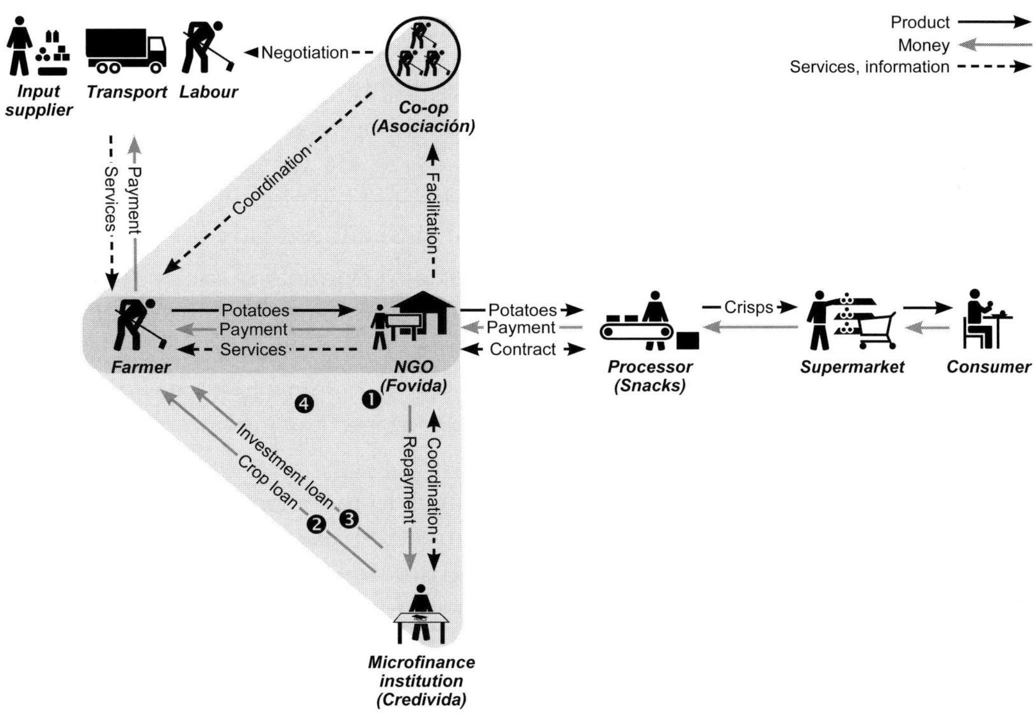

Figure 4.2 The potato value chain, 2001–7: Fovida is a chain actor with a coordinating role

the farmers did not sell the potatoes to other buyers. If the quality was below specification, the company would not accept delivery. This contract gave Fovida a lot of responsibility and a lot of risk: it became a major actor in the value chain (❶ in Figure 4.2).

Snacks advised Fovida on what varieties it required and the production technology to use, quality requirements, and so on. The Fovida technicians considered these recommendations, adapted them as appropriate, and informed the Asociación.

Fovida and the Asociación calculated the production costs per hectare, and estimated the yields. They took this information to Snacks, and with Fovida's help, the farmers negotiated a price directly with Snacks that gave them an estimated profit margin of 15%. When agreement had been reached, Fovida signed the delivery contract on behalf of the farmers.

As a result of these arrangements, the farmers started growing potatoes for Snacks. Now, from December and April each year, 85% of each farmer's land is planted with Capiro or Canchan potatoes. During the rest of the year the farmers continue to grow vegetables such as tomatoes, lettuce and carrots, as they had been doing before.

From field to supermarket shelf

The potato value chain was organized like this:

1. Fovida coordinated the supply of seed potatoes (from producers in the Andes) and other inputs to the farmers in time for planting.
2. Each member of the Asociación planted seed potatoes at different times, following a planting schedule designed by two Fovida agronomists working on the project full-time. The Asociación coordinated the planting and harvesting by the farmers to ensure a regular supply of potatoes ready for processing.
3. Throughout the growing season, the two agronomists made regular visits to advise farmers on production issues and help them solve problems. The average yield was 22–24 tons per hectare.
4. During the harvest season, the farmer whose turn it was to harvest that week hired workers to lift, sort and grade the potatoes. The agronomists were on hand to supervise this process.
5. The farmer hired a lorry to take the potatoes to the Snacks factory, which is located in an industrial zone in Lima, about 40 km away. The Asociación had an arrangement with a group of truckers to provide transport services, so each farmer did not have to negotiate prices individually.
6. At the factory, a Fovida staff member supervised the unloading and weighing of the potatoes. A new lorryload of potatoes arrived at the factory three times a day during the harvest season. One farmer could deliver an average of 24 tons of potatoes – two lorryloads – at a time.

Figure 4.3 *It is vital to grade the potatoes after harvest so they conform to the buyer's requirements*

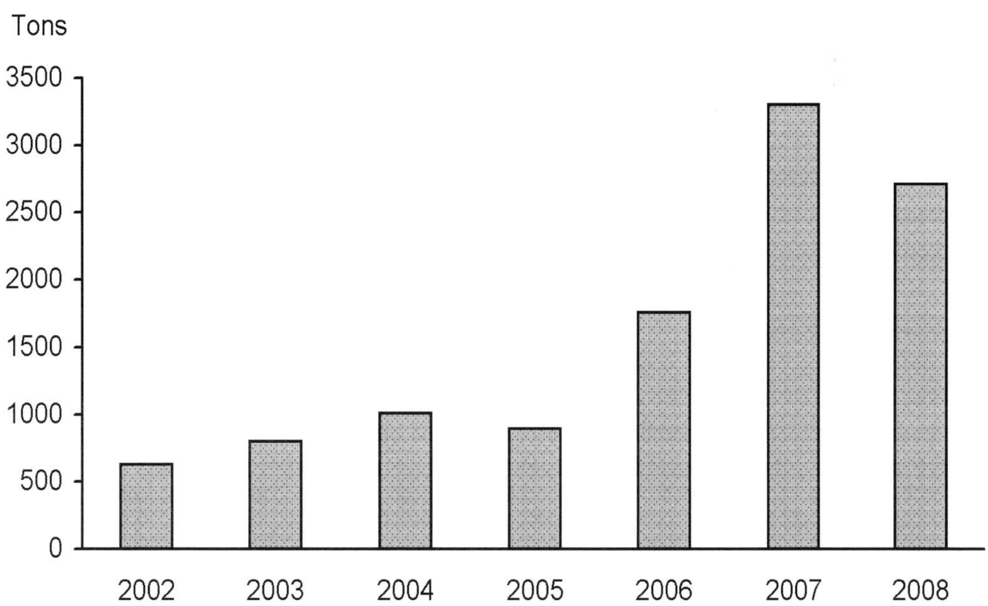

Figure 4.4 *Production of Capiro potatoes, 2002–8*

7 The Fovida staff sent the delivery data to the president of the Asociación. That meant the Asociación could monitor production, yields and deliveries, and was alerted to any problems that occurred.

8 Within 10 days of each delivery, Snacks paid Fovida for the amount supplied, and Fovida paid the farmers. Fovida deducted a portion of the credit (including interest) the farmer had received (see below), and a 2.6% fee.

9 Snacks processed the potatoes into crisps, packaged them and delivered the finished product to supermarkets, distributors and wholesalers.

This arrangement has worked well. Production grew from 629 tons in 2002, to 2,713 tons in 2008 (Figure 4.4).

Arranging credit

The farmers needed money to pay for seed and other inputs, as well as to cover other expenses while they waited for the payment at the end of the season. Fovida explored various potential credit providers, but found either closed doors or unattractive conditions. For example, they turned to Agrobanco, a state bank that finances agricultural chains at low interest rates. This bank was willing to provide loans to individual farmers, but it wanted the Asociación to provide hard collateral, which it did not have. The farmers also heard that Agrobanco's procedures were lengthy and bureaucratic, and its loans payments were often too late to be useful. So the Asociación and Fovida decided it was not practical to pursue this option.

What was the alternative? Without credit, the chain could not function. Peruvian banks and microfinance institutions could not offer the loans as required by the farmers, so Fovida had no other option than to look for a solution inside its own organization. Fovida runs a credit programme called Credivida (Box 4.1), which provides loans to urban entrepreneurs and had only limited experience in rural credit. Fovida was reluctant to involve Credivida because of the potential complications and conflict of interest (with the same organization providing credit

Box 4.1 Credivida

Credivida is a Fovida programme that has several years of experience providing credit to small urban entrepreneurs. Although it is part of Fovida, it is managed as a supervised entity, has its own accounting system, accumulates reserve funds as required by the Peruvian banking authorities, pays taxes on interest, and has achieved full economic sustainability.

Credivida's financial products include working and investment capital and short and long-term loans. It has a team of 8 professionals and operates in two of Peru's 25 regions. In June 2009 it had a loan portfolio of 3.8 million soles (about US$ 1.2 million).

More information: www.copeme.org.pe and www.mixmarket.org (search for Fovida)

as well as technical assistance). But there was no other option. Credivida agreed to tailor-make a financial product to suit the needs of the potato growers. It was willing to do this because in Snacks, the farmers had a firm buyer with a guaranteed price, and because Fovida was managing the chain, provided technical support to the farmers and handled all the payments.

With its long experience in urban credit, Credivida knew that if small entrepreneurs invested their money well, they would be able to honour their debts. It

Table 4.1 Credivida's loan terms for Asociación potato growers in Peru

	Crop loan (❷ in Figure 4.2)	Investment loan (❸ in Figure 4.2)
	Product and financial flows	
Purpose	Financing of potato growing for sale to processing industry	Finance for investments in farm equipment
Amount	US$ 2,000 per hectare	Average $6,000 for each producer
Period	6 months No penalty for early repayment	Up to 3 years with repayments every 4 or 6 months No penalty for early repayment
Disbursement	Direct to the farmer in 2 or more instalments	All at once
Repayment	Lump sum, including interest, after delivery of product to buyer	Every 4–6 months depending borrower's income flows
Interest rate	2.8–3.2% per month in local currency	1.2–2% per month in US dollars
Transaction costs	None	None
	Risk management	
Securitization	Borrower must own the land, but land not used as collateral	Hard collateral (such as house), particularly for higher amounts
Liability	The borrower	The borrower
	Information flows	
Information required to apply	Personal documents Characteristics of the family: number of members, ages, activities Number of hectares owned Area planted to potato	Items in previous column, plus: Plan on how to use the money Budget for item to be purchased
Information required during season	A field credit officer visits the farmer once a month to check on progress	A field credit officer verifies that the investment has been made
Time lag between application and payment	2–3 days	1 week on average

believed that the same would be true for farmers in a value chain like that supplying potatoes to Snacks.

Crop loans Credivida designed a standard credit line, personalized for each farmer (❷ in Figure 4.2, first column in Table 4.1). Standard items include the period of financing (6 months), the maximum amount of loan per hectare ($2,000), the interest rate (2.8–3.2%, depending on the amount borrowed) and the items to finance (seeds, fertilizer, pesticides, transport). This standard package could be adjusted according to the amount borrowed, the number of hectares, the possibility and amounts of partial payments, and the guarantors for the loan. The only collateral required was that the farmer had to own the land, or could get the owner sign in his or her stead (the land itself was not used as collateral). Credivida provided 100% of the loan directly in cash. The credit was made available when the farmers needed it to buy inputs.

Repayment terms Potatoes can be harvested all at once (or two or three times if the farmer grows enough to make multiple deliveries). So lump-sum repayments, including the accumulated interest, were scheduled after the expected harvest dates. Snacks paid Fovida, which deducted the amount of the loan and paid it to Credivida, before forwarding the balance to the farmer.

Interest rates Credivida has to cover its own costs. Its interest rates are an average of those charged by other microfinance institutions. The more a farmer borrows, the lower the interest rate it charges. Credivida accepts early repayments without applying any penalty.

Investment loans Credivida also designed an investment loan package to enable farmers to purchase equipment or other items. These loans were for up to 3 years, with repayment every 4–6 months. Hard collateral was required for these loans – typically the borrower's house, depending of the amount of the loan (❸ in Figure 4.2, second column in Table 4.1).

The credit scheme worked well because Credivida had a triangular relationship with the farmers and Fovida (❹). Though the farmers had no hard collateral, as required by the local banks, Credivida considered the farmers to be attractive clients, thanks to their integration into the potato chain. Credivida knew that the farmers would generate sufficient cash flow to repay the loans by supplying potatoes to Snacks. The risks were relatively low because the farmers were closely supported by Fovida. The solid relations in the value chain replaced the need for hard collateral, thereby making it a viable operation for Credivida.

Making the chain sustainable

The potato chain was working well and benefited all the stakeholders. But the system was not yet sustainable: it depended on the support from Fovida. Over time, as production and trade kept growing, the value chain evolved towards more business-like relations. As this happened, new triangles were built between Fovida, the farmers, Snacks and Credivida.

Fovida–farmers

By 2001, Fovida had become a major actor in the chain: it was organizing inputs, production and post-harvest activities; and it was guaranteeing the supply and quality of the produce. It was providing these services for free, using funds from Novib, a Dutch donor agency.

It was necessary for Fovida to take on these roles in order to get the value chain started, but this role was not sustainable for an NGO in the long term. Fovida believes that farmers should learn how to operate in the real market, so should pay the full cost of the services they receive. So Fovida needed a strategy to hand over responsibility.

In 2004, the Novib project came to an end, and a government-funded development project covered some of the extra costs. But not all: Fovida made it clear to the farmers that they would have to pay part of the costs. It started deducting a service fee of 1.31% from the payments to the farmers.

The government funding ended in 2007, so Fovida had to charge more. After long and sometimes difficult discussions, the Asociación and Fovida agreed that the service fee would be 0.028 sole cents per kilogram weight sold (approximately 3.78% of total sales). Reaching this agreement was possible only because Fovida was transparent about its costs: the agronomists' salaries, transport, administration and overhead. The producers realized that they would have to take on more tasks in the chain in order to keep Fovida's commission low.

In 2008, Fovida's fee covered 85% of the costs of its services; the NGO paid the remaining 15% out of other funds.

Because they are now paying more for the service, the farmers have become more demanding: they want more frequent visits from the Fovida field agents and want the process of billing and payment to be speeded up. This puts pressure on Fovida to improve the services it offers.

Input suppliers–farmers

When it first established the chain, Fovida arranged for suppliers to provide inputs to the farmers' group, rather than to individual farmers. This made sense at that time: the inputs would be cheaper when bought in bulk. But the farmers have decided to start buying from their input suppliers as individuals rather than as a group. There are three reasons for this:

- **Staggered planting** The farmers plant and harvest potatoes at different times in order to comply with Snacks' delivery schedule. Some farmers did not need the inputs right at the beginning of the season, but would have to take out a loan to pay for them then. That meant they would have to pay all the accumulated interest even though they had not used the inputs.
- **Operational problems** Some farmers complained that the group did not store seed and other inputs properly, or that they had not been provided the

correct amounts. Others did not pick up their supplies on time, leading to storage and management problems for the group.

- **Disagreements about suppliers** The farmers did not always agree where to source the inputs. Some complained that the suppliers chosen by the group's management provided inferior inputs.

These changes are shown as ❺ in Figure 4.5: the input suppliers now link with individual farmers, rather than with the Asociación.

Credivida–farmers

Throughout the years, the farmers built up a good credit history; only very few defaulted on their loans – usually if the potatoes did not attain the quality required. As Credivida became familiar with their production system and financial status, it expanded its loan services to other produce, including crops where Fovida has no involvement. It also started offering credit for investment and home improvements, with repayments scheduled for when the farmers tend to have cash. These loan products have performed very well: the farmers do not want to risk their potato loans by defaulting on one of the other loans.

Fovida's technical services and Credivida's credit facilities are run by different people and are operated independently of each other – though the farmers know they are really part of the same organization. That means that farmers may receive technical services even if, for whatever reason, they do not qualify for credit; or they may get credit although they do not use the technical services.

Credivida registers its loans with one of the most important central credit-reference bureaus in Peru, which other financial institutions can consult. In this way the farmers are building up a credit history and can apply for loans from other banks.

New triangles

In 2008, 6 years after the first contract between Snacks and Fovida, the company agreed to include the farmers' group as a formal partner in the agreement. Instead of a supply contract between Fovida and Snacks, there is now a tripartite contract between the company, the NGO and the farmers' group. In the new agreement, Fovida has stopped being a direct actor in the chain. The farmers have taken over full responsibility for producing and delivering the potatoes to Snacks in accord with the quality standards as agreed upon. Fovida is now responsible only for providing technical and business support to the farmers, in return for its service fee.

Fovida continues to manage the financial flows in the value chain. Fovida receives the payment for the product from Snacks, deducts its own service fee, repays the loan from Credivida, and then transfers the remainder to the farmers as a net payment for their produce.

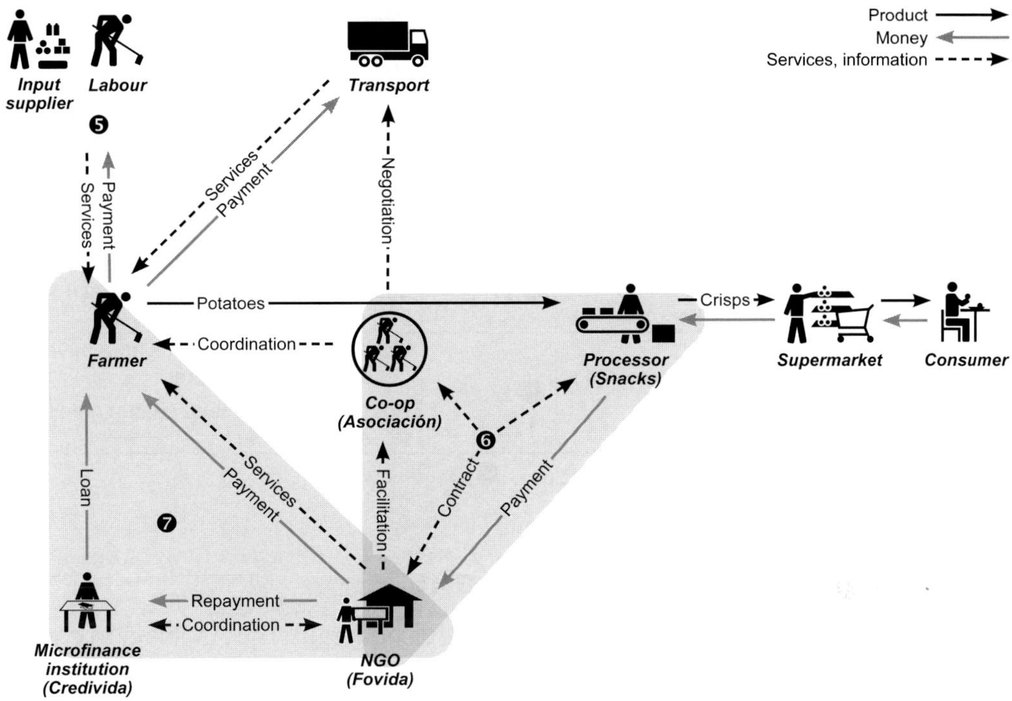

Figure 4.5 The potato chain in 2008: two triangles

These changes mean that there are now two triangles in the potato chain:

1 A triangular relationship between Snacks, Fovida and the farmers (❻ in Figure 4.5) in which the farmers produce and deliver potatoes, Snacks buys and pays for the produce, and Fovida provides technical, financial and business support services

2 A triangular relationship between Credivida, Fovida and the farmers (❼ in Figure 4.5) in which Credivida provides loans to the farmers, the farmers invest the loans to produce potatoes, and Fovida collects the payment from Snacks to repay the loan and pay the farmers.

Costs and risks

The potato chain involves several risks. The risks are mitigated because the chain actors work closely together in their triangular relationships (Table 4.2). To ensure they get good-quality seed, the producers make advance contracts with certified seed growers. Rising input prices and low yields are another problem; farmers overcome this with the support of Credivida's technical experts. Together, they monitor the crop and adjust the farming system if needed. Defaults on repayments are prevented by giving Fovida control over the financial flows in the value chain.

Table 4.2 Risk analysis of the potato chain in Peru

	Source of risk	Risk level	Mitigation strategy
Production	Poor seed quality	●○○	Advance contracts with certificated seed growers
	Input prices	●●○	Modify farming system to use less chemical inputs
	Low yields	●○○	Technical assistance from Fovida
Price	Price fluctuations in the potato market	●○○	Agreement on prices before season begins, allowing the farmers to make sound profits (15%)
Default	Non-repayment of loan	●○○	Fovida handles payments from buyer and deducts loan and interest Borrowers know about central credit reference bureau

Risk levels: ●●● High ●●○ Medium ●○○ Low

Fovida steps out of the chain

In 2009, the agreement with the buyer was revised again: the Asociación contract eddirectly with Snacks, without Fovida being involved. The tripartite arrangement has been replaced by a bilateral contract between the farmers and Snacks. The farmers may choose to contract separately with Fovida to supply technical and business support services (they have to pay the full cost of these). Or they are free to make arrangements with other service providers if they wish.

This means that Fovida has moved from being a major chain supporter with triangular relationships with Snacks and Credivida, to a mere supplier to the chain. This is shown as ❽ in Figure 4.6.

Building and destroying triangles

Back in 2001, it was necessary for Fovida to become a chain actor, taking responsibility for delivering potatoes to the buyer. Also, to get the necessary production loans, Fovida had to build a triangle that involved the producers' group (the Asociación), a financial institution (Credivida), and itself (❹ in Figure 4.2). Without this triangle, the value chain could not have been established: alternative sources of finance were not available.

But this role was not sustainable for an NGO: Fovida needed to move away from being a chain actor. It did this first by creating a new triangle involving Snacks, ensuring that the farmers negotiated directly with the buyer, with Fovida in a supporting role (❻ in Figure 4.5).

Now that the chain is functioning, Fovida can reduce its involvement further. From 2009 onwards, Fovida has become a mere supplier of services. The farmers have got used to the idea of paying for these services, and are rightly demanding value for money. The farmers may decide to go elsewhere for these services, rather than hire Fovida.

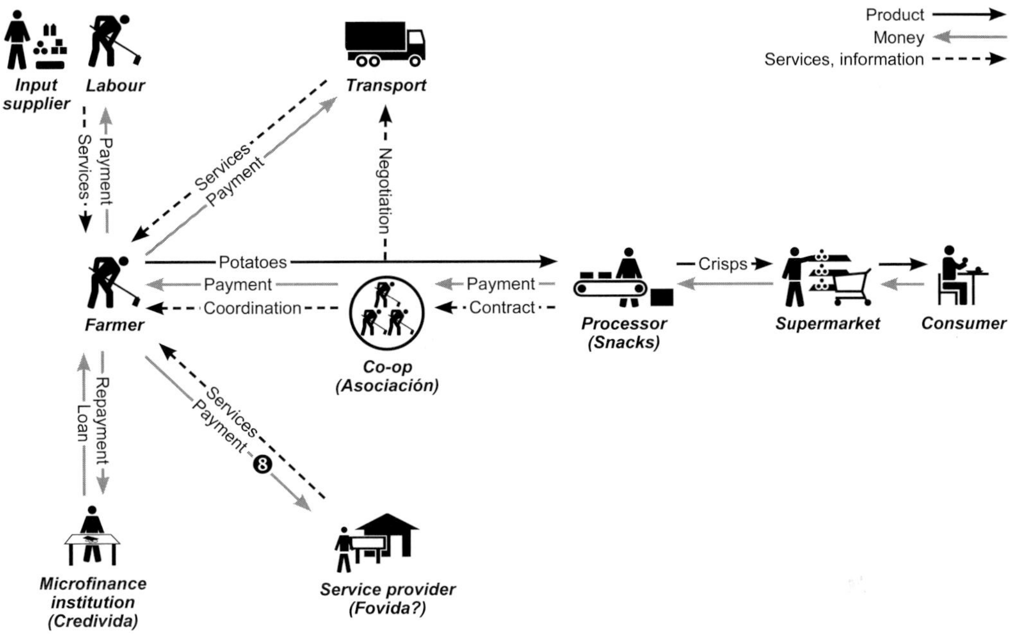

Figure 4.6 The value chain in 2009: farmers' group is responsible

As Fovida is no longer responsible for handling payments and deducting credit, there is no triangular relationship with Credivida either. The farmers have now built up their own credit histories and can get financing on their own. They may decide they no longer need Credivida's services: they can go to another financial institution if it offers a more attractive deal. Both triangles have disappeared (Figure 4.6).

Empowering farmers

These changes are entirely in accord with Fovida's and Credivida's mission. As an NGO, Fovida aims to empower the people it serves. That means building up a value chain until it can run independently, and Fovida can withdraw. The same is true for Credivida: while it is run on commercial lines, it is part of Fovida, so shares its overall goals of graduating its clients to mainstream commercial service providers.

Fovida empowered the farmers in various ways. It trained them in production techniques, business management, negotiation with the buyer (and with Fovida itself), and financial management. The farmers had to assume the costs, so they quickly learned to get value for money: they began to demand tailor-made financial services to suit their individual needs. To reduce their costs, they decided to contract their own part-time coordinator, and take on more responsibility for organizing the chain, planning production, and arranging for seed supplies, skilled harvesting workers and transport.

Impacts

The players have benefited in various ways from the potato value chain.

The producers have moved from being an unorganized set of individual growers who sold low-value products in an unsystematic way, to a skilled, organized group that supplies a high-value product, at a fixed price, to a sophisticated value chain. They have a strong relationship with the buyer, with prospects for the future. They have increased the area planted from zero 36 ha in 2002 to 170 ha in 2008, and total output from 630 tons to 2700 tons in the same period. Improved production technology has boosted their yields from 15 tons/ha to 23 tons/ha. They have an assured income, spread throughout the harvest season because of the staggered planting schedule.

The Asociación and its members are now familiar with financial institutions and contractual rights and obligations. They have developed their business skills: for example, they have learned of the need to analyse costs before negotiating a deal. The farmers have developed their skills for dialogue and negotiation, so are now more able to play an active role in the market. They have learned the importance of working together in businesses that require large-scale production.

The farmers now have a credit line from Credivida that covers not only potatoes but also other needs. One-third have taken out Credivida loans for equipment, vehicles or house improvements. They have developed their own credit histories, so are more likely to find open doors when they approach banks for loans. Their incomes grew significantly in the first years because of the high price Snacks paid; the farmers want to continue the arrangement so they can keep this price advantage. Many have started renting more land so they can produce more potatoes. (This also has a negative effect, as the cost of renting land has risen.)

Service providers Service providers such as truckers, field workers and harvesting teams have more employment and have learned how to work with organized producers.

Credivida has learned how to evaluate credit for agricultural purposes and is pleased with the loan model it has developed. It now has two professionals working exclusively on agricultural credit, and the service is financially sustainable. It has already expanded this model to three other value chains. Not only that: it has also started serving farmers who are not part of a particular target chain, but who grow other crops or raise livestock.

Credivida now has a growing portfolio of around 170 active farmer clients in the coastal area. It has moved from being an almost exclusively urban institution to one that also serves rural areas. Urban micro-enterprise credit is a highly competitive field, so this new rural credit line spreads Credivida's risks and gives it a more balanced and efficient portfolio. In 2008, about $1 million, almost half of its disbursements, went to the farming sector.

Fovida is on the point of relinquishing its coordinating role in the chain. It has transferred the major responsibilities to the farmers' association, and has fulfilled its mission of building strong, sustainable farmers' groups who can generate wealth and buy in the services they need. It has avoided the common trap facing NGOs that work with value chains: with a clear exit strategy, it has made itself unnecessary for the sustainability of the chain. And it has a model for building value chains with poor producers that it can replicate in other areas and with other commodities.

Snacks The buyer gets a constant, reliable supply of potatoes of uniform quality. In the producers' association it has a strong business partner that has shown that it can deliver the produce regularly without problems. Because Snacks is now assured of a regular supply of potatoes, it can use its production facilities more efficiently and can keep crisps on the supermarket shelves throughout the year. Customers are less likely to switch to other brands or products if they do not find the particular type of crisps they are looking for.

In 2008, Snacks started marketing a new product: crisps made from a multicoloured native potato variety, grown by smallholders in the Andes. The value chain works in a similar way to that in the coastal area, though is not yet self-sustaining. Snacks is interested in developing new products following a similar model.

Consumers Peru's many middle-class consumers can now find their favourite crisps on the supermarket shelves all year round.

The chain as a whole The value chain is now more efficient. Production facilities are idle for less time, and there is a constant flow of products in one direction and money in the other. The total amount of Capiro and Canchan potatoes grown and traded has increased from zero before the initiative, to over 2,700 tons in 2008. Information flows are transparent. The chain is governed by clear rules in which all chain actors have a say. It is sustainable: as long as Peruvians want to eat crisps, and as long as the actors in the chain remain competitive, there will be a demand for the potatoes produced in the Chillón valley.

Challenges

Monopoly buyer It is risky for the farmers to be so dependent on a single product and a single buyer. Fovida recognizes this, and has tried to link the farmers to other buyers. But the farmers prefer to maintain their relationship with Snacks.

New tasks Until now, Fovida has been handling the payments and documentation that are required. The Asociación is new to this function. It needs to ensure that it can handle these tasks in a timely, efficient manner.

Tensions within the Asociación As illustrated above with the input suppliers, the interests of the group members are not always the same. The Asociación will have to find suitable people to take over the administrative tasks from Fovida, fix their salaries, pay the expenses, and absorb potential losses. Larger-scale members may wish to contract directly with the buyer instead of going through the Asociación.

Lessons

Market-driven approach The potential for agricultural development is greater if it is guided by the market. In the Chillón valley, Fovida first identified a market (Snacks) and product (potatoes). It then identified a group of farmers who could supply this market, trained and organized them to do so, negotiated contracts and made the financial arrangements. This is an unusual approach for an NGO, which often start with a group of farmers, help them produce more or a particular commodity, then start looking for a market.

Initial investment Such projects require an initial investment, especially when working with a poor population. It was necessary for Fovida to help get the farmers organized, build their capacity and provide technical advice at first. After an initial period when the services were free, Fovida had to start charging fees and get the farmers to realize they would have to pay the full cost after a certain period. For financial services, the farmers had to pay the full cost right from the start.

Transparency Even if a donor is supporting the intervention, it is important that the NGO makes the full costs clear to beneficiaries (in this case, the farmers). That will force the NGO team to build low-cost models that the beneficiaries can take over when the NGO phases out. The buyer also had to be transparent in its operations, for example in controlling quality and avoiding hidden costs.

Covering transaction costs Managing the chain has its own costs: coordination, training, supervision, handling documentation and payments, and so on. The chain must be profitable enough to generate enough money to cover these costs. Also, the chain must have enough scale to cover for these costs. Otherwise there will be no way to manage the chain in the future when the NGO withdraws.

Integral interventions Fovida (technical, organizational and business support) and Credivida (credit) worked closely together to ensure that the chain operated smoothly. Without such cooperation, the chain may not have been as successful.

Empowerment For Fovida and Credivida, empowering the producers was an important goal. They kept this goal in mind throughout, even though Fovida had to take on more responsibilities in the chain than is normal for an NGO. Because it had a realistic exit strategy, it has been able to hand over this responsibility to the producers themselves.

More information

Luisa Santur Alberca, *Fovida, luisa@fovida.org.pe*

www.fovida.org.pe

Milk helps revive a microfinance institution in Bolivia

THIS IS A STORY of how a technically bankrupt microfinance institution managed to get back in business by restructuring its portfolio towards value chain finance.

Sartawi Foundation is a Bolivian microfinance institution focusing on small-scale farmers and rural micro-entrepreneurs. Established in 1990, by 2002 Sartawi had developed a portfolio of $5 million, of which 60% was for rural clients. Its main service was providing short-term loans for small farmers. But many borrowers defaulted – an average of 30%, and in 2003 half of them failed to repay. From 2002 onwards, Sartawi had a negative balance sheet (it had more debts than assets), and in 2005 it was on the verge of bankruptcy.

But by 2008 Sartawi was back in business. Its defaults were down to just 3% on a portfolio of $2.2 million. How did it achieve this? Did it abandon rural lending and focus instead on more profitable urban clients?

No – in fact, just the opposite occurred. Sartawi's portfolio became more rural, and its average loan size decreased. So rather than abandoning small farmers, Sartawi increased its work with them, while at the same time improving its own operations. What was the secret?

Sartawi developed a new financing model: one based on strategic cooperation in the value chain. This new model was pioneered together with Save the Children, an international NGO, in the milk value chain.

The traditional milk value chain

Although 60% of the population of Bolivia, the poorest country in South America, live below the poverty line, the economy has grown strongly in recent years (6.5% in 2008). The growing urban middle class has rising purchasing power. They like to shop in modern stores and are willing to pay higher prices for good quality. That goes for milk, too. Urban consumers look for fresh, safe milk in supermarkets rather than buying from street vendors.

This demand is supplied by Bolivian milk processors as well as imports. The two biggest firms that supply milk and other dairy products to the capital, La Paz, are Pil Andina and Delizia. These two firms had a total capacity of 60,000 litres of milk a day, but fell 24,000 litres short: they could not source enough milk from the farmers in the surrounding highlands, the Altiplano.

Why not? Let us look at the situation of the dairy farmers.

Typical dairy farmers on the Altiplano have just three cows that produce only 9 litres of milk a day, depending on the season (6 litres in the low season, 15 litres in the high season). Many of these farmers also raise sheep as well as cattle, and grow potatoes, barley, alfalfa and quinoa for home consumption. The farmers graze the cows on their own land, and forage is scarce in the dry season, leading to lower milk production. The farmers can get veterinary services but cannot afford to buy additional feed. They have no information about how the milk market works.

The farmers keep 10% of their milk for home use, and make cheese from the remaining 90%. They sell this direct to consumers at weekly rural markets nearby (Figure 4.7).

There are three main reasons the farmers do not increase their milk output and the market demand remains unsatisfied.

- Farmers' revenues are so low that they have no way to upgrade and expand their production. Farmers make on average \$2 a day by selling cheese, and another \$3 from selling other foodstuffs. That equals \$1 a day per person for a family of five - a case of extreme poverty. The farmers have no savings to invest and no incentive to increase production because the demand for cheese in the rural markets is limited, and prices are low.
- Dairy processors have no system to collect the milk from the producers. The roads are bad and there is no cool chain (milk collection centres, cooled tankers, etc.). The farmers would have to go themselves to the dairy plant to deliver small amounts of milk - but the dairy is a 3 or 4 hour bus ride away. The farmers are not organized, and it would be unprofitable for the dairy to collect milk from many individual smallholders.

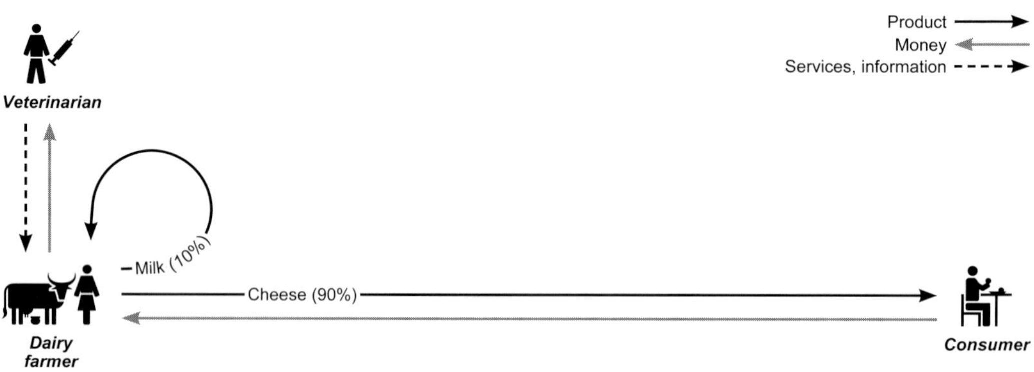

Figure 4.7 Previous value chain for milk

- To produce more, the farmers would have to buy higher-yielding cows and feed them with purchased feeds. They cannot afford to do either. No financial institutions are willing to lend money to individual farmers without collateral or a firm market. It is a vicious circle, a poverty trap.

Figure 4.8 Location of Sica Sica and Patacamaya, Bolivia

Intervening in the market

In an analysis of the Bolivian milk market, Save the Children found that the country produces 287 million litres per year, but consumes 460 million litres (the deficit is imported from Argentina and Chile). It saw that dairy farmers had an opportunity to increase their incomes if a way could be found to boost their production and build an efficient system to bring the milk from the highlands to the dairy plants. Small dairy farmers had a competitive advantage because the cold temperatures in the Altiplano mean their milk has a high fat content - a characteristic highly appreciated by dairy processors.

Convinced that the farmers could satisfy the demand, Save the Children met with the executives of several dairy companies to see if they were really willing to buy from small-scale producers. Its strategy was based on the following elements:

Finding a buyer The market analysis showed that the Bolivian milk market is dominated by three companies. The two biggest, Pil Andina (which has an 80% market share) and Delizia, showed strong interest in Save the Children's initiative. They were aware that the initial volume would be low, but they could see the potential of the value chain approach. Save the Children agreed to start a pilot with both firms.

Organizing the milk producers In 2006, Save the Children helped a group of 60 farmers from 12 communities in Sica Sica, some 115 km from La Paz, form the "Association of Milk Producers from the Carmen Altiplano" (Aplec). In nearby Patacamaya there already was a group of 30 farmers from 4 communities which was also incorporated into the project (Figure 4.8). Some of the farmers agreed to supply Pil Andina; others agreed to deliver milk to Delizia.

Technical assistance The technicians of Save the Children initiated "farmer field schools" to teach the farmers about animal hygiene, health, nutrition, breeds, herd management, milk collection and delivery, and business management. Save the

Children is not responsible for verifying the quality of the milk the farmers deliver; this is done by the dairy, which reports any problems to Save the Children.

Increasing production In the beginning the farmers in the two groups were producing a combined total of just 80 litres of milk for sale a day. That is less than 1 litre per producer: they were risk-adverse, looking how the project would develop, and in the meantime they continued producing cheese. This was, of course, far below the volume that the companies needed. The producers still needed better animals and feed to increase the volume of milk they produced. So buying new animals and feed was crucial to make the value chain work.

Pil Andina, the larger company, was able to provide its farmers with feed. But Delizia is a smaller firm and could not afford to do so. So Save the Children needed to involve a partner that could provide financial resources to the farmers who supply Delizia. Otherwise there was no way to increase production and keep the company interested in working with the small farmers.

Searching for a financial partner

Finding a financial institution to finance the milk producers was far from easy:

- **Banks** required hard collateral to securitize loans: proof of income and an inventory of assets. Few farmers could fulfil these requirements. Moreover, the banks had no branches in the countryside to serve the farmers.
- **Microfinance institutions** offered loans but at very high interest rates (25% a year) and most worked only in urban areas, so their financial products were not tailored to the farmers' needs.
- **Development finance institutions** (a type of NGO that provides financial services on a non-regulated basis) were approached, but most did not respond positively. Some worked exclusively with urban women. Others worked with farmers, but were not prepared to put in their own funds; they would administer the portfolio for Save the Children but charge high rates.

Sartawi Foundation: The ideal partner

But one development finance institution, Sartawi, seemed a good partner: it had a rural focus and a specific mission to serve small farmers and rural micro-entrepreneurs. Sartawi's managers saw that the dairy farmers were enterprising, were receiving technical assistance, and were producing high-quality milk. At the same time, there was a large market demand for milk. Sartawi was also encouraged by the fact that the producers' clients, Pil Andina and Delizia, were leading dairy companies in Bolivia, with a consolidated market. The milk farmers had contracts with both companies, guaranteeing the dairies would buy milk at an agreed price as long as it met the quality standards. This contract could act as "soft collateral" to reduce Sartawi's risk.

Figure 4.9 Farmers use their profits to buy food and pay for their children's education

Save the Children's proposal was fully in line with a new business strategy that Sartawi had developed to restructure its operations. Sartawi had had a couple of tough years and was on the verge of bankruptcy. Though it focused on the countryside, it had had little knowledge of the value chains it financed or the actors in those chains. Rural communities were far from its head office in La Paz, complicating the process of assessing loans and collecting payments, and raising operating costs and increasing its risks. As a result, Sartawi's loan appraisal was inadequate, resulting in a very high-risk agricultural loan portfolio and a 30% default rate. Plus, producers' collateral, such as farm equipment, was of no use to Sartawi, so in cases of default it had no value.

Sartawi needed a new strategy for agricultural loans, one which would improve credit risks and lower costs. It needed "legs" in the rural marketplace - a partner with experience and knowledge of the communities - to reach the farmers. Sartawi realized that Save the Children could be this partner, and agreed to join the pilot project. For Sartawi this would consolidate a new form of working that it had been exploring since 2005. The new model aimed to reduce risks through the following elements:

- **Preferred supply relations** between the producers and a purchaser.
- The purchaser is a **lead firm** in its sector.
- The producers supply a **large amount** of the raw materials required by the purchaser.

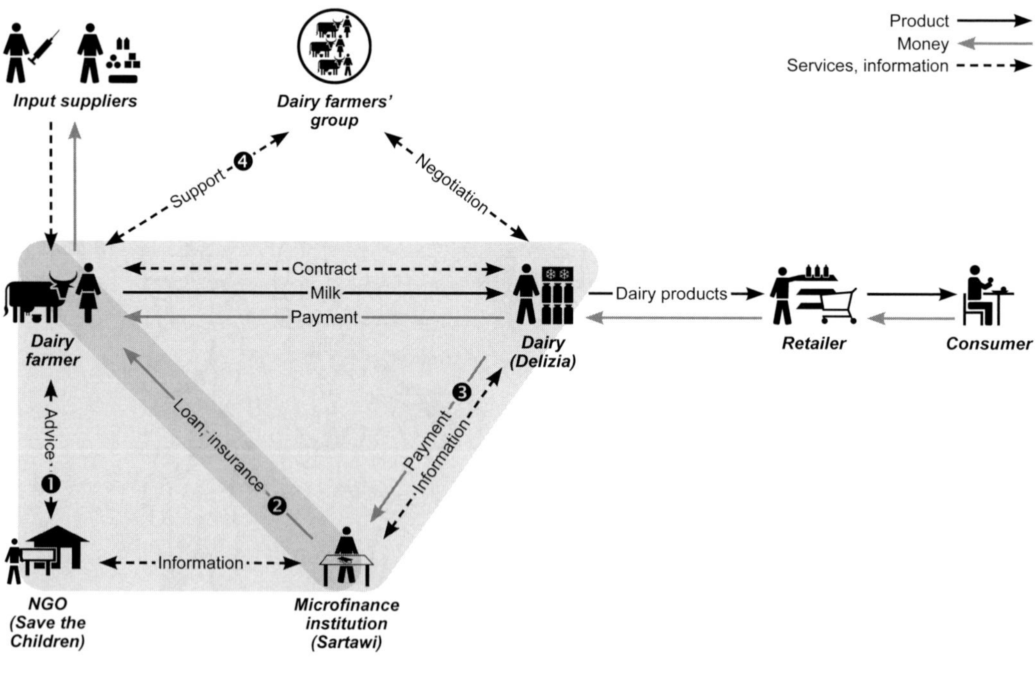

Figure 4.10 The new milk value chain

- In the overall market for the raw materials that the farmers produce, **supply is lower than demand**, so the purchaser is likely to buy from the producer.
- When paying the farmers for their produce, the **purchaser retains part of the produce value** to pay back the loan, which reduces risks and transaction costs.
- The loans are combined with **micro-insurance** in order to mitigate risks by transferring these to the insurance company. This reduces the interest rates on the loan.
- The model of cooperation must be attractive for producers, purchaser and financier so that it can be **scaled up or replicated**.

The new dairy value chain

The resulting value chain includes two strategic alliances, one for technical assistance, and another for financing:

- **Save the Children + Sartawi + farmers** Save the Children provides training and technical assistance to the producers (❶ in Figure 4.10), and gives Sartawi the information it needs about the production system to provide financial services to the farmers. Save the Children pays for its services with donor funds.
- **Sartawi + Delizia + farmers** Sartawi provides individual loans and micro-insurance to the farmers so they can purchase improved animals and feed, among other things (❷ in Figure 4.10). The farmers produce milk under contract for Delizia, which pays them every 2 weeks, retaining a share to repay their loans from Sartawi (❸). To manage this scheme, Sartawi and Delizia share the account information for each farmer, including the repayment schedules, monitoring sheets, payment receipts and bank transfers.

The farmers' groups are important supporters of this new chain. They organize their members, help them prioritize their needs, negotiate prices with the dairy, support their members' loan requests, and ensure that loans are repaid (❹). They are not directly involved in their members' commercial and financial relations, which remain responsibilities of the individual.

Sartawi's financial services for the milk farmers

In 2006 Sartawi began providing loans to the dairy farmers of the Altiplano. The average loan was less than $1,000. The farmers used the loans to improve their stables, buy high-quality feed for the cows, and to buy more productive animals. Table 4.3 gives details of these loans.

Sartawi follows these steps when appraising a loan application:

Table 4.3 Details of Sartawi's loans to dairy farmers in Bolivia

Product and financial flows	
Purpose	To finance dairy farmers' operations and investment capital needs
Amount	\$100–3,000 Average per farmer less than \$1,000 Total loan portfolio (pilot programme) of \$100,000 in milk value chain
Period	6 months–2 years, depending on farmer's payment capacity and purpose of loan
Disbursement	Disbursed directly to dairy farmers by Sartawi in La Paz
Repayment	Payments by bank transfer by Delizia
Interest rate	Up to 16% per year in US dollars
Transaction costs	1% administration fee (one time, payable at time of disbursement)
Risk management	
Securitization	Up to \$2,000 without hard collateral Securitization is based principally on sales contract between dairy farmers and Delizia Collateral can also be provided in the form of livestock, equipment or a guarantor
Liability	Individual farmer
Information flows	
Information required to apply	Identity card Status in central registry of financial system Copy of sales contract with Delizia and other necessary guarantees Registration of membership of the Dairy Farmers Association Basic plan on how loan will be used: purpose, cash flow projections, repayment, etc. (prepared with Sartawi's assistance) Inspection of farm by Sartawi staff Information about previous season and other references
Information required during season	Sartawi makes inspection visit to farmer at least once every 2 months Save the Children and Delizia have constant local presence and inform Sartawi about problems that could put repayment at risk
Time lag between application and payment	After requirements fulfilled, maximum 5 days to approve and disburse loan

1 **Application and evaluation** The milk producer applies for financing from Sartawi. Sartawi evaluates the application according to its guidelines and procedures to determine the borrower's capacity to repay.
2 **Validation and risk management** Sartawi checks with Delizia on the amount of milk the farmer has sold. This helps assure Sartawi that the farmer will be able to repay the loan. Additionally, Save the Children provides qualitative information about the client (family, actual capacity, risk, behaviour, etc.).
3 **Disbursement** Once Sartawi approves the application, it disburses the funds to the farmer.
4 **Repayment** Delizia has an agreement with the farmer and Sartawi to retain a portion of the amount it owes the farmer for his or her milk. It transfers this amount to Sartawi to repay the farmer's credit. This greatly reduces Sartawi's overhead and operational costs since it does not have to collect money from each individual borrower. As a consequence, Sartawi can lower its costs, becomes more competitive, and can offer more attractive interest rates to farmers.

The loans for the dairy farmers bear an interest rate of between 13 and 16% a year (preferred clients pay lower rates). Sartawi obtains loans at 3% interest from the PROFIN Foundation, a microfinance institution supported by Swiss and Danish development agencies. Sartawi adds 7% to cover its operating costs, 3% provision for defaults, and between 0 and 3% profit, making 13–16% interest in all. Sartawi's normal rural operating costs are 14%, but are only 7% in this case because Delizia and Save the Children take over parts of the due diligence, monitoring and enforcement. Sartawi's profit margin is intended to let it raise its own capital for operations.

Mitigating risk

Many of Sartawi's risks are mitigated because of the farmers' relationships with Save the Children and Delizia. These two organizations are aware of the borrowers' needs and production capacity, making it easier for Sartawi to assess the purpose of the loan and the borrower's ability to repay it. Save the Children and Delizia field staff monitor the farmers and can detect problems quickly. They share information about individual farmers' accounts. The directors of the NGO, the dairy firm and the microfinance institution meet regularly to discuss their cooperation.

Table 4.4 summarizes the risks to the loans and Sartawi's strategy to mitigate them.

Sartawi mitigates the risks of accident and death of the borrower and damage to property by offering micro-insurance to borrowers. It charges a small fee for this coverage on top of the interest costs of the loan. This service is managed through an agreement with an insurance company based in Bolivia:

Table 4.4 Risk analysis of the milk value chain in Bolivia

	Source of risk	Risk level	Mitigation strategy
Production	Drought	●●○	Save the Children provides additional feed
	Livestock disease	●●○	Save the Children and Delizia provide information and technical assistance on disease
	Quality and quantity	●●○	Save the Children and Delizia provide technical assistance on inputs, productivity, etc.
Price	Price fluctuations	●○○	Purchase guarantee at fixed price from Delizia
Market	Lack of market demand	●○○	If Delizia cannot buy milk, Pil Andina will do so
Default	Side-selling	●○○	Farmer signs contract with Delizia Save the Children, Delizia and farmers' association supervise deliveries
	Non-repayment of loan	●●○	New loans will be frozen for Association members
Currency	Exchange rate	●○○	Sartawi debts and farmers' loans are in US dollars
	Inflation	●●○ long term ●○○ short term	Assessment of farmer's payment capacity is conservative to allow for inflation
Other	Transport blockades, political unrest	●●○	Delizia to extend its network of cooled tanks in the are to hold milk for a while
	Accident or death of borrower, damage to property	●●○	Micro-insurance

Risk levels: ●●● High ●●○ Medium ●○○ Low

- **Insurance in case of borrower's death** This protects family members against inheriting the debt. The family also receives $200 to cover funeral expenses. This scheme also covers borrowers against total or permanent handicap. The fee is 0.09% per month of the outstanding debt.
- **Life insurance** Family members receive $2,000 if the borrower dies naturally, and $4,000 in case of accidental death. The annual fee is $18.
- **Multi-risk insurance** against material damage that might affect the farm enterprise. Coverage includes damage to property (except land), machinery, and mobile equipment (except cars). The annual fee is $12 for coverage worth $1,500 and $18 for $10,000.

Benefits

Farmers The farmers have a new source of revenue (milk rather than cheese) which diversifies their risks. Their numbers have grown: the programme began in 2006 with 90 farmers producing just 80 litres of milk a day. By 2009, 340 farmers were producing 3,000 litres a day. The average daily income per farmer has increased from $0.30 to $3 a day. This tenfold increase results from a 60% rise in productivity (due to balanced feed) and an increase in average herd size from three to five. Farmers are producing less cheese: they still consume 10% of their milk at home, and sell 80% to the dairy, leaving only 10% to make cheese. Their income from cheese has fallen accordingly.

The farmers in Sica Sica have registered their organization, Aplec, with the government and affiliated it to the association of milk producers of the Department of La Paz. This gives the farmers greater negotiating power with the diary companies.

Their relationship with the dairy company has given the farmers soft collateral that makes them eligible for credit from financial institutions. Micro-insurance is teaching the farmers how to care for the future of their enterprises and their families.

The farmers have learned about milk production and financial management, gaining knowledge and skills that lead to higher yields and competitiveness. With these skills, they will ultimately be able to carry on without Save the Children's support (the technicians of Delizia will take over), while maintaining the progress they have achieved - access to loans, higher incomes, greater food security and strengthened relationships with other chain actors.

In essence, the farmers escaped from the poverty trap: they have become skilled actors in an integrated value chain with long-term business relations; this increases their access to financial services, which in turn enables them to further invest and grow.

Dairy company Delizia has diversified its milk suppliers and can rely on a regular, growing supply of high-quality milk from a reliable set of suppliers.

The chain as a whole Consumers can buy more milk. The supply of cheese has gone down, but demand for this was low anyway. The farmers have higher incomes, so can spend more, boosting the local economy.

Finance institution Sartawi has reduced its transaction costs (and thereby its interest rates) because Delizia and Save the Children have taken over part of the credit appraisal, monitoring and payment enforcement. The low-interest loan from PROFIN enables Sartawi to provide loans to farmers at attractive rates.

The value chain finance model has allowed Sartawi to better manage and clean up its loan portfolio. The average loan amount disbursed fell from $2,072 in 2005 to only $869 in 2008, and the average size of loans in its portfolio decreased from $1,980 in 2003 to $977 in 2008. That shows that Sartawi is succeeding in serving smaller, poorer clients, deepening the fulfilment of his mission. From 2005 to

2008, the number of clients rose slightly from 2,125 to 2,277. The total portfolio size has shrunk accordingly, from $5 million in 2002 to $2.2 million in 2008. Milk loans account for 20% of this.

The quality of Sartawi's loan portfolio has increased dramatically. The default rate has fallen from 17% in 2005 to 3.5% in 2008. Sartawi has introduced this way of operating into value chains for peaches and apples, and now 94% of its loan portfolio is based on chains where relations are a key element. Not counting bad debts from the past, the default rate is now only 1.95%.

In line with its mission, Sartawi has increased its operations on the countryside. Its loan portfolio has become more rural, from 60% in 2002 to 75% in 2008. It has developed in-house expertise, partnership networks, and a line of tailored financial products, making it a leading player in the field of rural finance in Bolivia.

Challenges

Continuity Save the Children will finish its intervention in the milk value chain at the end of 2009. Seeing the project's profitability, Delizia has already engaged the services of a dairy specialist, who now works together with Save the Children and who will continue with the technical assistance to the farmers when the NGO withdraws. There is so much competition among dairies for the milk that they need to provide technical assistance in order to keep their producers loyal.

New insurance In order to reduce risk for farmer and financers alike, it will be important to develop micro-insurance for livestock deaths, as well as for crops.

Replicating the model in other sectors Based on its positive experience in the milk chain, Sartawi wants to further develop and replicate the model to other value chains, such as soybeans, potatoes and vegetables. The model clearly reduces the credit risk for the financer. It is hoped that other financial agents will also apply this model.

Lessons

Win–win cooperation Strategic collaboration can create a win–win situation for all parties involved. The dairy farmers increased their incomes, Delizia satisfied its unmet demand for milk, the financial institute increased its agricultural loan portfolio and decreased the credit risk, and Save the Children was able to work towards its goal of decreasing chronic malnutrition among Bolivian children, through better incomes for farming families. These achievements were possible because strong, trust-based relationships were developed ("soft collateral") between all four parties, enabling clients without hard collateral to gain access to credit, and creating sustainable relationships among the parties.

More information

Navil Agramont Akiyama, *Save the Children Bolivia, nagramont@savechildren.org.bo, nagramont@hotmail.com, tinyurl.com/yj6qycr*

Marcio Oblitas Fernández, *Sartawi Foundation, marcio.oblitas@sartawi.org, marcio_oblitas@hotmail.com, www.sartawi.org*

Delizia Compañía de Alimentos S.A., www.delizia.com.bo

Creating a new dairy value chain in India: Reliance and BASIX

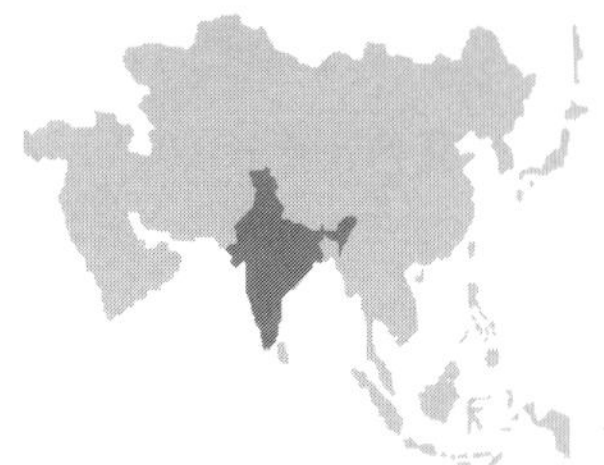

EVERY MORNING, A QUEUE forms in the dusty street outside the village house. Men, women, small children and young people all wait patiently for their turn. Some chat with their friends; others are impatient to get away to school or household chores.

A doctor's surgery? No: the queue is moving too quickly for that. And everyone is carrying a jar or bottle of milk. One man has a big metal bucket, carefully covered to prevent the dust from getting in and the contents from spilling. A child is clutching a bottle that cannot contain more than half a litre.

Looking past the queue as it snakes in though the door, it's possible to see what is happening inside. Each person steps up to a desk, where there is some electronic equipment run from a car battery on the floor. The man behind the desk uses the

Figure 4.11 The collection centre accepts even tiny amounts of milk

equipment to test some of the white liquid in each container. Each person then tips the milk into a gleaming steel churn standing on a set of scales. A readout shows how much milk has been poured in, and the man hands the person a printed chit of paper showing how much milk he or she has delivered, and how many rupees he or she is owed.

The same people will be back in the evening, and the following morning, and the evening after that, each bringing half a litre, one litre, three litres of milk. For this is one of the most remarkable dairy operations in the world: one that buys milk from thousands of individual small-scale buffalo owners and sells it through supermarkets in India's booming cities.

Dairying in Mahabubnagar

Mahabubnagar is one of the poorest districts of Andhra Pradesh, a state in southern India - so poor that every year, several thousand residents migrate to other parts of the country in search of a better life. Dairying is important here: 40% of rural families depend on their buffaloes for milk to drink and to sell. An average family earns US$ 100 a month, with milk accounting for some 50–60%. A family might have one or two animals, which they milk once a day. Most buffaloes are indigenous breeds; they yield only 2–2.5 litres of milk a day. Buffaloes are traditionally the major livestock in the area: they provide 90% of the milk in the district; cows produce the rest.

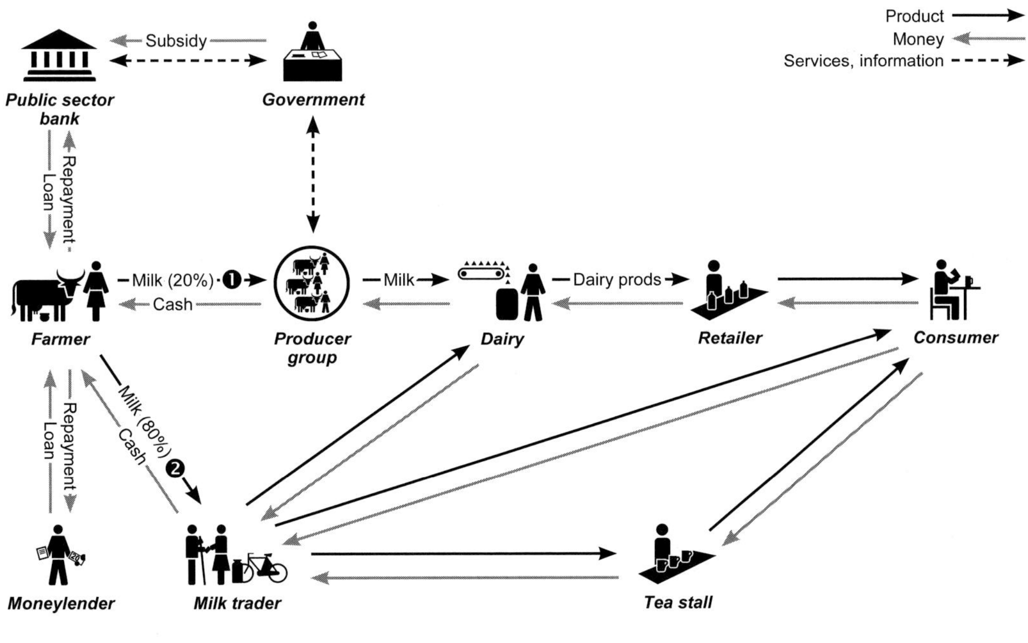

Figure 4.12 An inefficient milk value chain

Figure 4.13 Location of Mahabubnagar in India

The government has built a lot of infrastructure to support the dairy industry in the district. It has supported the farmers to work together in groups, and it established dairies that buy buffalo milk from the farmer groups to produce pasteurized milk, yoghurt and other dairy products (Figure 4.12).

Unfortunately this government-run system is inefficient and underused: it handles only 20% of the surplus milk produced (❶ in Figure 4.12), and the dairies run at only 30% capacity. This is because the chain is not well organized: the farmers are paid only once a month, there are inadequate controls to prevent adulteration of the milk, and the dairies are subject to political patronage and interference. Several government development schemes provide subsidized loans to the farmers, but they often fail to reach the poorer farmer groups like minorities and lower caste individuals. As a consequence, many farmers are chronically short of money and heavily depend on informal money lenders.

For these reasons most farmers sell their milk to private traders (❷ in Figure 4.12). The traders offer better prices and pay every 15 days, sometimes even in advance. Because of this, the traders buy 80% of the surplus milk. They visit the producers each day on a moped to pick up the morning's milk. Some sell their milk to the government-run dairies; others sell milk to tea stalls and other local retailers, while others sell raw milk direct to consumers.

But this privately run system has its own problems. The milk is often diluted or adulterated, and there are no checks for health or quality. This is a great concern for many Indian consumers. As India's middle class is booming, the demand for fresh, high-quality milk products is increasing rapidly. Rather than buying unlabelled milk from a tea stall in the open air, these Indian consumers want to find a wide range of dairy products, nicely packaged and presented, in coolers in their supermarket. They are willing to pay a higher price for the milk in exchange for being assured of its quality and safety.

Spotting a business opportunity

Reliance Fresh is a modern supermarket chain operating in this new market segment of urban consumers with higher purchasing power. The company is part of the Reliance Group, India's largest industrial holding. To secure a steady supply of fresh milk for its retail outlets, the Reliance Group established a specialized

Box 4.2 Franchising

All three tiers of the collection system – village pooling points, cooling centres and dairies – are run on a franchise system. The owner must provide the building and staff; Reliance provides equipment to the pooling points and cooling centres, and trains the franchisees and staff how to use them. A network of Reliance field staff supervises and supports the three tiers.

This model reduces Reliance's capital and running costs, since Reliance does not have to build all its own infrastructure and employ lots of staff. It also spreads the risk, since each franchisee is responsible for delivering the product to the next stage in the chain, and bears the risk if something goes wrong.

dairy firm in 2006: Reliance Dairy Foods Ltd. It supplies dairy products to the Reliance Fresh chain of supermarkets, as well as to other urban retailers.

When setting up its supply chain, Reliance chose not to source milk from the government-run dairies. The firm was not confident that these dairies could assure the quality and quantity required, and it was concerned about the payment terms and the danger of political interference. The company also decided not to use the existing private traders: it wanted to extend opportunities to others.

So Reliance decided to build its own milk chain. It designed a supply chain consisting of three tiers:

1 **Village pooling points**, where the farmers can bring their milk on a daily basis, even in very small portions (❸ in Figure 4.14)
2 **Bulk milk cooling centres**, where the milk from various village pooling points is collected and cooled (❹ in Figure 4.14)
3 **Dairies**, where the milk from various cooling centres is processed, packaged, and distributed to the market outlets (❺).

Reliance figured that, rather than controlling the full supply chain by itself, the best way was to train selected local villagers and provide them with the necessary infrastructure and technical support, so they could operate the pooling points and cooling centres as independent business people. This is a so-called franchise model (see Box 4.2).

Building an alliance

Reliance started pilot operations in two areas: in Mahabubnagar, and in Punjab in the north of the country. But Reliance realized that it could not do the job alone. The company was new and it had no experience in organizing rural villagers and training them to become part of a supply chain. It needed a strong partner to ensure that the village end of the chain functioned well.

Box 4.3 The BASIX group

The BASIX group is a private company which promotes livelihoods for the rural poor. The group includes both for-profit and non-profit units that provide financial services, technical assistance and consulting services. It currently works with 1.1 million households in 20,000 villages across 17 states in India. Among other things, it provides microfinance to dairy farmers. It has been working in Mahabubnagar District since 1997. Reliance was interested in building on this experience, and BASIX wished to replicate its successes.

In 1999–2001, BASIX worked with a milk-chilling centre belonging to the government-run Andhra Pradesh Dairy Development Cooperative Federation at Wanaparthy, a town in Mahabubnagar District. This chilling centre has a capacity of 10,000 litres/day, but was handling only 1,250 litres. BASIX offered technical assistance and support services to improve the producers' management and productivity. It loaned a total of Rs 6 million ($120,000) to 500 producers to buy 600 milking animals. The repayments were routed through the dairy co-op federation, saving the need to collect payments from individual producers. Milk production rose significantly, and within a year, the chilling centre was procuring over 10,000 litres a day. The net income of a typical producer household rose by Rs 5,700 ($115) a year.

In 2001, responsibility for BASIX's financial and technical assistance operations in Mahabubnagar District was handed over to Krishna Bhima Samruddhi Local Area Bank (**KBS Bank**), a BASIX subsidiary. This is a private, profit-oriented local bank, incorporated in 1999, and granted a license in 2001. It offers savings and other financial services to the rural poor in Mahabubnagar district Pradesh and two districts in neighbouring Karnataka. It currently has 12 static and two mobile branches in the three districts, serving 1,585 villages with over 110,000 customers

More information: www.basixindia.com

In January 2008 it signed an agreement with KBS Bank, a subsidiary of BASIX (Box 4.3), to organize and support milk producers in Mahabubnagar and to build up the Reliance value chain. Reliance pays BASIX a service fee of Rs 0.40 for each litre of milk procured from its customers.

The Reliance/KBS Bank team surveyed villages in the district to discover their milk production potential. They appointed local people to run the village pooling points and cooling centres. They set up the infrastructure in each location, and trained these people how to manage the equipment and run the service as a franchisee. They supported the farmers to elect one person in each village to act as a joint account holder to receive and manage the milk payments on behalf of all the farmers.

Establishing this infrastructure and training the villagers to run it took 6 months. By then the new milk value chain was reality (Figure 4.14).

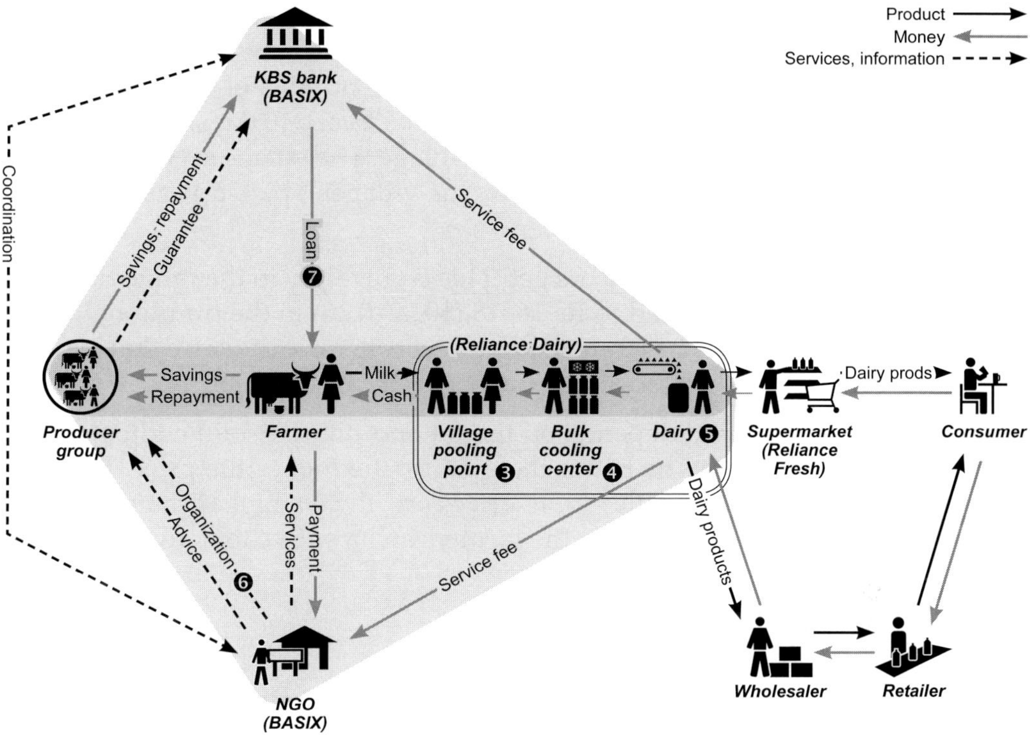

Figure 4.14 The Reliance/BASIX value chain

Collecting milk: The Reliance chain

Village-level pooling points

This is where farmers bring their milk each morning and evening (❸ in Figure 4.14). Each lot of milk is tested and weighed before the farmer pours it into the waiting steel churn. The average farmer delivers 3 litres of milk per day to the village point. Some deliver tiny amounts – as little as half a litre. Some 2,040 producers currently supply Reliance with milk at 40 village pooling points.

To be sustainable, each village pooling point must have at least 50 milk producers who supply a minimum of 150 litres of milk a day. The pooling point is run by a "village service provider", selected at a meeting of the villagers. This person is a franchisee (Box 4.2): he or she must provide a room where the producers can bring their milk. The franchisee is paid a commission of 3.75% of the value of the milk delivered each month.

Reliance provides each village pooling point with testing equipment, a machine to remove air bubbles from the milk, an electronic scale, steel milk cans, and a data processing unit with dedicated software. The average investment per village pooling point is $1,480.

When the producer pours the milk into the churn, he or she is given a printed receipt showing the weight, fat content, non-fat solids (which shows whether the milk has been diluted) and acidity (which reflects the number of bacteria), as well as the price. These data are recorded electronically. Every 10 days, Reliance pays the money into a joint bank account for each village. The village service provider and the joint account holder, appointed by the villagers, pick up cash from the bank and disburse it to the producers.

Reliance fixes the price the producers get. This is currently in the range Rs 20–26 ($0.40–0.52) per litre, compared to Rs 16–18 ($0.32–0.36) in the traditional value chain. This price difference has forced other buyers to raise the price they pay for milk by Rs 2 ($0.02) per litre, so all producers in the area benefit.

The price paid by Reliance depends on the fat and non-fat solid content of the milk each farmer delivers: the better the quality, the higher the price. Reliance pays extra for good-quality milk, so its prices are always higher than what the government dairies pay. That gives the farmers an incentive to deliver a quality product.

Bulk milk cooling centre

The cooling centre (❹ in Figure 4.14) is situated in the main village in a cluster of settlements. It serves 10–12 village pooling points within an average radius of 10–12 km. Reliance pays for the milk to be brought to the cooling centre by a three-wheeled auto-rickshaw.

Reliance has set up four cooling centres in the region served by BASIX, plus several more in other areas. They also operate on a franchise basis: the owner gets a commission from Reliance of Rs 0.30 ($0.60) per litre of milk delivered (1.3% of the value). Reliance provides all cooling equipment: a compressor, a bulk cooling unit, a solar water system, testing equipment, standard milk cans, chemicals, and a computer. Reliance has invested around $14,300 in equipment for each cooling centre.

Each cooling centre can handle 2,000 litres a day. To be profitable, it has to operate at a minimum of 60% of its capacity. The current capacity is 83%.

Dairies

Reliance has negotiated with defunct or underused private dairies to give priority to processing its milk (❺ in Figure 4.14). It pays to transport the cooled milk to these dairies, which also operate on a franchise model: Reliance pays them an annual lease that depends on their capacity and location.

The dairies pasteurize and package the milk. The finished product is sent to wholesalers, supermarkets and retailers, from where it reaches consumers. Up to 25% currently goes to Reliance Fresh. Reliance organizes the flow of products in this chain.

Management information system

The entire milk supply chain is managed by a state-of-the-art management information system with easy-to-use equipment, a foolproof backup and uninterruptible power supply. This measures milk quality and amounts, calculates payments, and prints payment receipts and calculates the amount of milk procured from each producer, village and cooling centre.

BASIX's services

BASIX works in 40 villages (with the same number of pooling points) that supply four cooling centres. It promotes farmer producer groups in each village, provides agribusiness development services, and extends microfinance to the producer groups. KBS Bank provides loans, while BASIX provides agricultural and buisiness development services for KBS Bank customers.

Producers' groups

BASIX helps producers in each village form groups to improve their productivity and management abilities (❻ in Figure 4.14). Members of the producer groups are farmers who wished to expand and improve their dairying activities. The average group has 15 members. BASIX plans to form groups in all 40 villages; by January 2009, it had formed 17 of them.

These groups are the focus of the various services provided by BASIX.

Agribusiness development services

BASIX offers various business development services to members of the producer groups:

- **Improving productivity** Selecting stronger and more productive animals, advising on feed and fodder management, and buying inputs in bulk.
- **Disease prevention** Vaccinations against foot-and-mouth disease, regular deworming, and periodic visits by veterinarians to check on the animals' health.
- **Market linkages** Forming producer groups and providing institutional development services to help farmers supply milk to the Reliance pooling points.
- **Capacity building** Training on bookkeeping, accounting, management and help with formalizing the group.

These services are provided by BASIX field staff who visit the producer group members once a fortnight.

Financial services

Once a producer group is formed, it opens a joint account with KBS Bank and starts saving regularly (usually $1–2 per member each month). If it performs satisfactorily for at least 3 months, the bank will consider providing investment loans to the individual members (❼ in Figure 4.14, Table 4.5). Hard collateral is not required, but group members must all guarantee repayment of the loan. The bank can disburse the loan within 15 days of the application. The average loan is for the equivalent of $440 – enough to buy a buffalo.

Borrowers are responsible for repaying their loans on a regular basis. If the member falls behind with repayments, the group as a whole must pay up, so peer pressure is an important and effective way of ensuring repayment.

KBS Bank also provides savings services and livestock insurance coverage to the members of producer groups.

Risk analysis

By January 2009, KBS Bank had provided a total of $47,000 in credit to seven producer groups to buy new milking animals. These loans are equivalent to only 1% of the bank's total disbursements to the agricultural sector; they are not yet a major part of the bank's portfolio, but have great potential for expansion.

Normally, providing investment loans to farmers is very risky, because of the long repayment period and the uncertainty whether the investment will generate enough cash inflows to repay the loan. Therefore many financial institutes refrain from providing such loans in the agricultural sector. But the triangles of cooperation between KBS/BASIX, Reliance and the farmer groups (Figure 4.14) have helped to reduce many risks in the milk value chain. The key elements for risk reduction are (Table 4.6):

- The technical and business development assistance provided to the farmers by BASIX's fieldworkers.
- The commitment from Reliance to buy the milk and to pay premium prices for quality milk.
- The fact that the farmers operate in groups through which they support and control each other and build assets through regular savings.
- The sophisticated management information system put in place in the milk chain which allows for close monitoring of the cash flows and repayment of the loans.

The triangles of value chain cooperation have turned the investment loan into a financial service that is attractive to the farmers and at the same time viable for KBS Bank. The interest rate is 18%. This is 3% less than what KBS Bank charges for equivalent loans to other borrowers in the agricultural sector. It can charge less because the risks are smaller and the transaction costs are lower (because the bank is dealing with a group, not lots of individuals).

Table 4.5 Investment loan terms for Reliance milk producers

(❼ in Figure 4.14)

Product and financial flows	
Purpose	Financing purchase of milking animals (e.g., graded Murrah buffaloes and crossbred cows) to increase milk production
Amount	Individual loans are $440 per farmer Estimated loan portfolio for 40 villages = Rs 12 million ($240,000)
Period	18 months
Disbursement	To individual borrower in producer group
Repayment	Cash-flow-based repayment up to 36 months by producer groups
Interest rate	18%, reduced gradually as principal is repaid
Transaction costs	2% loan processing fee
Risk management	
Securitization	No hard collateral Group responsible to repay loan if individual member defaults. Peer pressure if a member delays repayment 80–85% of asset value is financed by bank
Liability	Each group member is liable for other members' payments
Information flows	
Information required to apply	Details of producer group formation and members Current status and dairy experience of each member Bank account details of group's savings Minutes of monthly meetings Record of group's financial transactions Repayment history of group members Resolution by group to apply for loan Estimated loan amount required for each individual and group as a whole Feedback from Reliance village service provider Counterparty check by bank customer service representative
Information required during season	Monthly savings of group Reports of monthly meetings and resolutions Milk quantity delivered to Reliance each month by each group member Payments made by Reliance each month to group members Livelihood service provider feedback on animal health
Time lag between application and payment	15 days from date of application by bank staff
	continued...

Table 4.5 (continued) Investment loan terms for Reliance milk producers

Other financial products	
Livestock insurance	Loan customers have livestock insurance to cover their newly purchased milking animals Premium = 3.9% of amount insured + Rs 100 ($2) service charge + Rs 13 ($0.25) service tax
Savings products	Group members can put monthly savings in KBS Bank accounts. This helps them manage the surplus from dairying and other activities. The level of savings may influence the future credit support.

When analysing the loan risk, KBS Bank uses the following sources of information:

- **Producer group status** The bank analyses the track record of the producer group and its members' creditworthiness. The group must be functioning smoothly, have records of its formation and subsequent activities, and its members must have been saving with the bank for at least 3 months. The borrowers must be approved by the group as a whole, and they must already own at least one milking animal.
- **Information from the supply chain** KBS Bank field staff and Reliance staff are in frequent contact with the franchisees and the farmer group members. They share the information from the management information system and they keep each other informed of problems encountered in the field.

Benefits

The value chain intervention has brought several benefits to the three key stakeholders: milk producers, KBS Bank/BASIX and Reliance, as well as to the chain as a whole.

Milk producers Over 2,000 farmers now sell milk to Reliance: an average of 50 farmers at each village pooling point. As a result of the BASIX technical support and KBS Bank's financial services, they have increased their sales from an average of 2 litres to 3 litres a day. They get around one-third more per litre than if they were to sell to other buyers. This means that they have more than doubled their daily income from milk.

Producer group members now have easier access to credit at affordable rates. They can get credit 3 months after they form a group, and only 15 days after applying for a loan. Some 105 producers from the seven producer groups have obtained credit. All have taken out livestock insurance, so their livelihoods are protected in case of death or accidents.

The farmers can also get regular technical advisory services for a modest fee to help them improve their production.

Table 4.6 Risk analysis of the dairy milk chain in India

	Source of risk	Risk level	Mitigation strategy
Production	Survival of high-bred buffaloes	●●○	Advise on breeds suited to local conditions Livestock insurance coverage
	Diseases	●●○	Vaccination, deworming, health checks, visits by field staff, government support
	Low fat content of milk	●●○	Promote azolla feed as supplement and better feed and fodder management
Price	Temporary price rises by local vendors	●●○	Strengthen relationship with milk producers through groups and training
Market	Adulteration of milk	●○○	Transparent quality parameters, price incentives to produce good quality, education on consequences of poor quality
Default	Farmer does not use loan to buy milking animal	●○○	Peer pressure by group members; follow-up by field and bank staff
	Sale of animal	●○○	Regular monitoring by bank and field staff
	Animal death	●○○	Livestock insurance coverage and claim settlement

Risk levels: ●●● High ●●○ Medium ●○○ Low

KBS Bank and BASIX Of the 2,000 farmers selling milk to Reliance, 500 are now KBS Bank customers.

Seventeen producer groups with 400 members use agribusiness development services from BASIX on behalf of KBS Bank. They pay the full cost of these services ($9 per member per year). As of January 2009, the 17 producer groups had saved a total of $4,000, and were saving an average of around $500 per month.

The central government apex bank, NABARD, has approached KBS Bank to promote more such producer groups in the district.

Reliance Reliance procures about 7,500 litres of milk a day. This is 83% of its capacity – well above its 60% break-even. The milk is good quality: the fat content is 7.1% and the non-fat solids are 9.4%.

A viable business model has been worked out for the village service provider cooling point franchisees. The transportation costs are reasonable.

Reliance has scaled up this approach beyond KBS Bank's operating area. It is now procuring milk from 350 villages via 29 cooling centres and a higher-capacity milk-chilling centre in Andhra Pradesh.

A new high-capacity chilling centre, handling 15,000 litres/day, has been proposed in the KBS Bank operating area.

Chain as a whole The chain as a whole benefits: more milk, of better quality, is available to consumers. The chain is more efficient, and produces a higher value product than before. The example and competition of this chain motivates other players in the market (including the government) to improve their operations.

Challenges

Availability of grassroots-level staff It is difficult to find suitably qualified field staff such as extension workers who can provide the types of services needed.

Competition from subsidized credit The government sponsors credit programmes that offer low interest rates, 3–9 % a year. Farmers hear of these, so are reluctant to accept loans at commercial rates. In fact, the subsidized programmes are rarely available, so farmers are left without any loans.

Loan cancellations A recent government decision to forgive loans may discourage farmers from keeping up their loan repayments in future.

Effect on other milk traders In the villages where it operates, Reliance handles 50% of the milk sold. Total milk production has gone up, so more milk is being marketed. Many farmers continue to sell milk door-to-door in nearby towns, as well as to traders other than Reliance. The traders often have good political connections, so Reliance is eager to avoid creating local political problems by diverting too much milk away from alternative market chains.

Replication Reliance and BASIX have developed a viable approach to buying milk from very small-scale producers. BASIX plans to us this partnership as a model for dairying elsewhere in India, for example by reviving defunct chilling centres or procuring milk in new locations.

Lessons

Big business benefits the poor Reliance is a big, profit-oriented company. Its mission is to implement business models that work - not to help the poor. Indeed, angry shopkeepers fearful of losing their livelihoods have burned down Reliance supermarkets in some cities in India. Such incidents have forced some state governments to ban Reliance Retail (the parent company of Reliance Fresh). But the success of Reliance's involvement in the dairy industry in Mahabubnagar shows that profit orientation is not incompatible with the interests of the poor. The producers get higher prices than before. Reliance's milk procurement system is very efficient, and it produces a high-value product: good-quality dairy products, conveniently packaged, sold through supermarkets. To ensure a reliable supply of raw material, it is prepared to pay the producers a premium.

Paid services for farmers The farmers get low-interest loans, advice on business development, livestock insurance, and technical advisory services. These services enable the farmers to improve their management, raise their productivity, improve livestock health, and produce better quality milk. The farmers pay for some of these services directly; the rest is covered by a fee paid by Reliance.

This shows that the poor are able and willing to pay for services that support their major livelihood, where they get support from the rest of the value chain, and where the transactions are transparent.

Regular savings Small-scale farmers often have a problem saving on a regular basis. The system provides a mechanism for them to do so and inculcates a savings habit. This is in fact the basis for the whole chain intervention.

Commitment to the whole chain If large companies wish to develop the dairy sector, they need to go beyond merely seeking business opportunities. They must be committed to the whole value chain: they need to invest in the whole chain for it to function, and to collaborate with other actors in the chain.

More information

***Mendu Srinivasulu**, KBS Bank, sreenivasulu.m@basixindia.com, mendusreenivas@gmail.com*

***Beeranna Palyam**, Reliance Dairy Foods Limited, beeranna_palyam@rediffmail.com*

BASIX, www.basixindia.com

Fighting elephants by growing chilli in Kenya

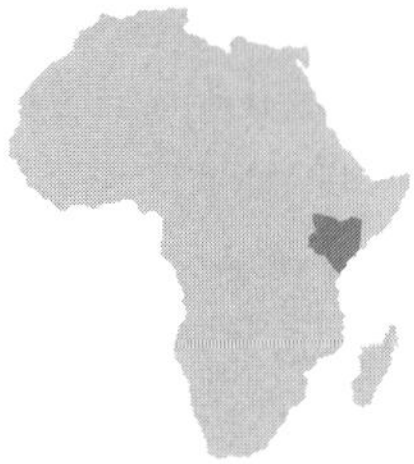

"Is there anything we can grow that the elephants will not eat?"

The farmers in Laikipia West district were fed up with wild animals eating their maize and beans. Invasions by elephants have increased over the years in the district, destroying property and trampling farmers to death. The Kenyan Wildlife Service does not compensate farmers for their losses, and wildlife is protected by law, so the farmers cannot kill the elephants that have made their lives unbearable.

Figure 4.15 Chillies are profitable – and act as an elephant repellent

Laikipia West is a semi-arid district in Kenya's Rift Valley, about 250 km north of Nairobi, with an average rainfall of 600 mm. Life here is tough even without elephants as neighbours. Various things make cropping and livestock-raising risky: recurrent droughts, infertile soils, cattle rustling, and disputes over grazing between livestock owners and farmers. Inputs are costly, and farmers find it difficult to get credit to buy them. Farmers in the area typically own 1–5 acres of land. Declining production leaves most farmers short of food, and people in the area have long depended on church and government relief.

In 2005, the Nakuru Region Inter-diocesan Christian Community Services (CCS) started a project in partnership with the Inter-church Organization for Development Cooperation (ICCO) to improve the income and livelihoods of people in the district. The team conducted a study to analyse local people's problems and come up with solutions.

CCS was already working with the farmers on other development initiatives, and had built up a good relationship with them. The farmers belong to producers' organizations of about 10–20 members each.

New members are free to join the group by paying a fee and monthly contributions. By registering with the authorities, the groups qualify for government grants as well as training on leadership, chilli production, business skills and financial management. CCS organized a series of meetings with these groups to discuss the findings of the study.

Growing chillies

The farmers chose a solution that had been suggested by the Kenya Wildlife Service - to plant hot chilli peppers. Elephants do not like the smell of the chillies, so leave the plants alone. Growing chillies around fields also protects the crops grown there. Chilli plants are resilient and easy to manage. They grow well in Laikipia West: they can withstand mild drought, need few inputs, and suffer from fewer pests and diseases than other horticultural crops. The average production per acre (0.4 ha) is 3,000 kg of fresh chilli. A high-value crop in demand throughout the year in local and export markets, it can be sold fresh or dried: a kilogram of fresh chilli fetches KSh 40 ($0.50), while a kilo of dried chilli is worth KSh 70 ($1) (chillies

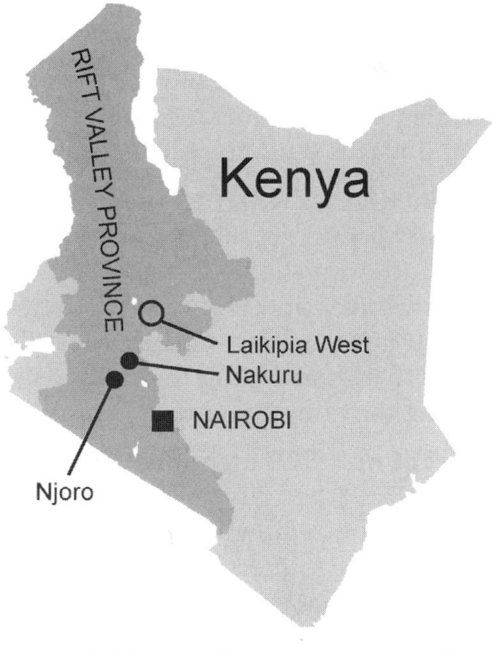

Figure 4.16 Location of Laikipia West district, Kenya

Table 4.7 Cost and income from growing chilli in Laikipia West

Activity/inputs	KSh per 0.25 acre (0.1 ha)
Seedbed preparation, nursery care, watering	1,000
Seeds (25 g)	200
Hiring land	1,000
Cultivation	2,000
Planting, manure	1,200
Pesticides, insecticides	550
Fungicides	400
Labour, chemical applications	1,200
Cultivation	750
Weedings (twice)	1,200
Harvesting (750 kg x KSh 5/kg)	3,750
Total costs	**13,250**
Gross income (750 kg x KSh40/kg)	**30,000**
Gross margin	**16,750**

lose about half their weight on drying). This is more income per acre than any crop the farmers had planted before.

Seed and other inputs Some 150 farmers bought seed at a good price (KSh 200 for 25 g of seed, enough to plant a quarter of an acre), but the first crop in 2006 did not yield well because of very poor germination and drought. Tests by a government laboratory confirmed the seeds were of poor quality, but despite admitting that it was at fault, the seed supplier refused to compensate the farmers.

Production The farmers sow two crops of chillies a year. They sow seed in a nursery bed in February, transplant to the fields 6–8 weeks later, and harvest in July and August. The second planting is ready for harvest during the short rains of November and December. After harvest, the farmers sort the produce and pack it into containers, ready for collection. The farmers prefer to sell fresh chillies rather than drying them: they lack storage facilities, and drying is difficult during the rainy season.

Marketing CCS identified various firms that buy chillies for processing or resale. Two of these are the main buyers of chilli in the area: Njoro Canning Factory (Kenya) Ltd. deals with dried chilli, while Frigoken buys fresh produce for export. CCS and the farmers discussed with these potential buyers issues such as varieties, likely quantities, sale conditions, modes of delivery and payment. The firms signed contracts with the farmer groups to grow the chillies. The firms fetch the produce from the farms, then check the quality before processing or packing.

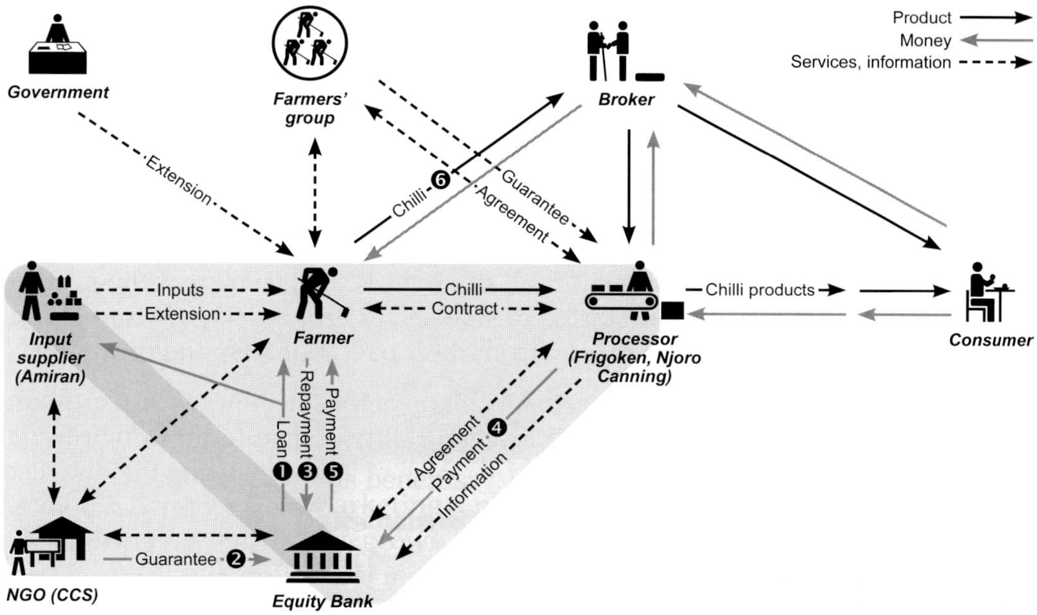

Figure 4.17 The chilli value chain in Laikipia West, Kenya

Financing the chain

In order to grow chillies, the farmers need loans to pay for seeds and agrochemicals (Table 4.7). They suggested Equity Bank as a potential source of credit. This bank, which provides financial services to the under-served, began as a rural microfinance organization in 1984 with only 10 farmers. It now offers banking services with a deposit base nearly KSh 12 billion ($150 million) and over 700,000 customers. Its typical rural client is a farmer who engages in some form of commercial farming. In general, farmers must have a savings account. Opening an account with Equity Bank is easy – it requires only KSh 400 (though account holders must have saved for at least 6 months before they can qualify for a loan). The minimum balance is small, and the bank is customer-friendly.

The bank was initially sceptical about lending to farmer groups: it doubted that chilli was a viable crop, and did not think the farmers were creditworthy or able to meet the bank's requirements. The farmers, for their part, were fearful of taking on a loan. To overcome these doubts, CCS guaranteed the initial loan (❶ in Figure 4.17) with a deposit of KSh 1 million ($12,500) (❷). The loan covered about 40% of the production costs; it was divided amongst the farmers in proportion to the area of chilli that each would grow. The farmers covered the remaining 60% of the costs.

Under an agreement between CCS, Equity Bank, the processor and the farmers' groups, the farmers start repaying their loans to the bank immediately, in instalments of KSh 350 ($4.38) a month (❸ in Figure 4.17). Within 3 weeks of receiving the chillies from the farmers, the processor pays the money into the group account at the bank, rather than to the individual farmers directly (❹). The bank deducts the remainder of the loan and interest, and pays the balance to the individual farmers (❺) (the group advises the bank how much each farmer has produced).

The failure of the first crop in 2006 put a major strain on this new relationship. The farmers were unable to repay their loans on schedule, and in desperation began to sell their chickens and cattle in order to do so. The bank's fears had become a reality, and CCS had to cover a shortfall of about KSh 200,000 that the farmers could not repay. Some farmers started to side-sell to brokers (❻) in order to prevent the bank from deducting money from their payment from the processor.

Everyone was discouraged by this setback. Many of the farmers gave up growing chillies. The producers' organizations lost credibility and suffered from internal conflicts when some members refused to repay their loans, insisting that the seed company compensate them for loss of earnings. The processors were disappointed by yields that were a far cry from what they had been promised. And despite the CCS guarantee, Equity Bank was disappointed by the failure of the enterprise.

Despite the disappointment, some of the farmers were still enthusiastic about growing chillies, and were confident enough in CCS to give it another go. The bank also realized that if the teething problems could be overcome, chilli growing would be viable because it was less subject to drought or wildlife damage than alternative crops.

A new seed supplier

In the next season, CCS linked 500 farmers to a different seed supplier, Amiran Kenya Ltd., and helped them negotiate a deal to buy seed in bulk and have it stocked at the local shops. This firm's seed costs much less than that supplied by Frigoken but is certified and is excellent quality: germination rates approach 90%. Amiran also supplies other inputs such as fertilizers, pesticides and fungicides, and trains farmers how to grow chillies and apply chemicals.

CCS arranged with Amiran and farmers who did not wish to get loans to buy inputs such as seed, fertilizer, pesticides and fungicides directly from Amiran stockists. The bank, on the other hand, gives farmers individual loans directly so they can buy inputs.

The partners overcame the initial problems and the value chain is now working smoothly. Everyone is clearer on their roles. The farmers now have a group account as well as individual accounts with Equity Bank. The individual accounts help them develop a savings habit, while the group account is used to buy in bulk and to distribute payments from the processor. In case a member defaults on a loan, the bank will deduct the amount from the group account.

The farmers are keen on keeping up with their repayments, and there are fewer defaulters - although still around 20%. To discourage the farmers from side-selling

Table 4.8 Equity Bank's loans to chilli farmers in Kenya

Product and financial flows	
Purpose	Financing production of chilli in Laikipia West District
Amount	Total loans to farmers in pilot project phase: KSh 600,000 ($7,500) Initial number of farmers (2006): 150 Current number of farmers (2009): 500+
Period	6 months
Disbursement	During March, through input suppliers Equity Bank also gives loans to individual farmers
Repayment	Farmers repay in monthly instalments of KSh 350 ($4.38) Buyer pays bank for produce 3 weeks after harvest; bank deducts outstanding amount and interest before paying farmers
Interest rate	10% per year
Risk management	
Securitization	Initial guarantee by CCS by depositing KSh 1,000,000 ($12,500) with Equity Bank Producer organization members now guarantee each other directly: in case of a default, the bank deducts the amount automatically from the group account
Liability	Farmer group guarantees loan repayments
Information flows	
Information required to apply	Group registration certificate Letter from Ministry of Gender, Culture and Social Services Farmer group must have a bank account
Information required during season	Occasional field visits to farmers by project staff and producer groups during the growing season
Time lag between application and payment	3–4 weeks

to brokers or other processors which are not part of the chain, the farmer organizations have signed a memorandum of understanding with the Njoro Canning Factory and Frigoken covering how much should be produced.

Table 4.8 summarizes the terms of Equity Bank's loans to the farmers.

Beginning with 150 farmers in 2006, the initial problems led to a halving in the number of farmers involved in the project. But encouraging returns in 2007 attracted 500 farmers to take part in 2008, and even more in 2009 (Table 4.9).

The chilli growers still have a long way to meet the market demand for the produce. The Njoro factory has capacity to handle at least 100 tons of dry chilli a year, while Frigoken require 30 tons of fresh chilli a week.

Table 4.9 Chilli production in Laikipia West, 2006–9

Year	No. of farmers	Total output (tons)	Processor	Issues
2006	150	4	Njoro	Bad seed Drought
2007	80	<4	Broker	Farmers needed immediate cash
2008	500	15–20	Frigoken, Njoro	Good season Some disruption due to political unrest
2009	>500	>500		Projected

Other actors

The brokers are informal traders who buy chillies from the farmers and pay on the spot, but at a much lower price (an average of KSh 25/kg for fresh chillies and KSh 40 for the dry product – just over half of the price paid by the processors). They in turn sell the chillies to processors or directly to consumers. These brokers have emerged as the value chain has developed. So far, only a few farmers have side-sold a small proportion of their chilli crop in this way, and this does not affect the loan recovery rates. So long as the value chain remains functional, with processors who pay good prices and financial arrangements which ensure reasonably fast payment to the farmers, side selling should not be a major problem. But the threat is always there to motivate all the actors to play their roles efficiently.

Business development service providers include the Department of Social Services, which trains the farmer groups in group dynamics, and the Ministry of Agriculture, which provides extension services.

Mitigating risk

Back in 2006, the chilli chain was built from scratch. The crop was new to the farmers and there were no established business relations. Because of the high risk of such venture, Equity Bank initially insisted on a loan guarantee from CCS, backed by a deposit of KSh 1 million ($12,500). This prudence was justified, as the crop failed due to drought and bad seed.

Over the years, the chilli chain developed and improved its performance. Seed supplies and yields have improved, the farmer groups have solidified through overcoming the initial problems, and the business relations in the chain have proven to work well. Now the bank no longer requires a guarantee. It relies instead on the mutual guarantee provided by the farmer groups, the savings in their individual and group bank accounts, as well as the fact that the farmers have contracts with buyers that pay much more than the private traders. In

Table 4.10 Risk analysis for chilli in Kenya

	Source of risk	Risk level	Mitigation strategy
Production	Drought	●●●	Harvest rainwater or runoff for supplemental irrigation
	Conflict	●●○	Initiatives for peace between crop farmers and livestock raisers
	Seed quality	●●○	Buy certified seeds from reputable company
Price	Chilli demand and supply	●●○	Organic production of chilli to expand market further Diversify (dairy goats, chickens, bees, etc)
	Side selling	●●○	Encourage farmer groups to sell entire crop to the agreed processors. Timely payments by processors
Default	Failure to repay loan	●○○	Group guarantee Processor pays bank, which deducts outstanding amounts before paying farmer

Risk levels: ●●● High ●●○ Medium ●○○ Low

other words, the performance in the value chain serves as soft collateral and has substituted the initial requirements for hard collateral.

Table 4.10 summarizes the risks in the chilli value chain.

Lessons

- Farmers are willing to trust institutions they have worked with for long periods, particularly if there is a faith-based element to the relationship. This trust can survive even serious setbacks.
- Regular reviews of performance by the different actors build trust, a joint sense of purpose and commitment. It helps if the actors have signed contracts and agreements to ensure responsible business practices (Box 4.4).
- Farmers and their groups manage loans better when they are trained in credit management.

Challenges and solutions

- **Side-selling** Farmers are tempted to side-sell their produce even though they have contracts to deliver to a certain buyer. But as the farmers build a more solid financial base, they are less in need of the instant payments that brokers provide.
- **Over-reliance on cash crops** Farmers (especially men) tend to increase their chilli production at the expense of crops the family can eat. As they improve

Box 4.4 Access to loans makes a difference

Joseph Njoroge woke up one morning to find his fence destroyed and his entire field of maize flattened by elephants.

"I was so disheartened", he says. "The maize and beans yields had really gone down anyway because of the recurring drought. Before, if I was really lucky, I could harvest 20 sacks of maize from the acre of land I leased. I could sell 10 sacks at KSh 700 each. My family and I would eat the rest. I would also plant beans between the maize for a yield of 2–3 bags. I could sell two at KSh 1,800 each and keep one for myself. The KSh 10,600 I would earn once a year was not enough to cover the cost of inputs and labour, let alone meet my needs.

"I looked for someone or a bank to help me with a small loan so I could buy fertilizers and implements. When CCS came to our area in 2005, my neighbours and I were eager to share our problems as well as listen to their ideas. They were willing to start projects on honey and chilli. Chilli is a hardy crop, requiring minimal inputs, has high yields and can be used to chase away the elephants. We got to learn later that if we mixed elephant dung with the chilli and lit the mixture, the smoke would irritate the animals and keep them away from the farms, protecting our other crops.

I was very sceptical about planting the chilli because I did not believe I could make any money from it. And anyway, I could not feed my family on chilli if I failed to sell the crop. CCS told us that they were willing to guarantee loans for us from Equity Bank. This was an answer to our prayers. We had our suspicions about financial institutions because we had heard frightening stories of people losing their property for failing to repay their loans. I had already tried to get loans and I knew that bank processes were tedious and out of reach for me. But CCS encouraged us and said they would support us through the process. They also advised us to use our self-help groups to ensure accountability.

"The agreement with the bank was that I would repay KSh 350 each month for 6 months to offset the loan of KSh 7,500 (plus 10% interest) and pay the balance on selling the chilli. The advantage of chilli is that I can harvest it at least twice a year. I was able to harvest 450 kg from my half acre of land. I sold 365 kg of fresh chilli at KSh 40 per kg, and 85 kg at KSh 70 per kg to Njoro Canning Factory. I made about KSh 20,550 in 4 months. In between, I would sell small amounts to brokers for my daily financial needs. From the other half acre, I still had some maize and beans to feed my family. I now had the confidence to borrow more money. The following year, I borrowed KSh 20,000 directly from the bank. They trusted that I would be able to repay the amount on schedule, which I did. I have now applied for a loan of KSh 33,000, which is yet to be approved."

their farming practices and gain experience, they get higher yields, allowing them to continue planting food crops so that they stay food-secure.

- **Danger of drought** Drought may force vulnerable families who borrow money into a worse financial situation than before, since they may have to sell valuable items in order to service their loans. Working with a bank introduces a culture of saving, which cushions such farmers and enables them to diversify their enterprises.
- **Scaling up** The large number of farmers wanting to start growing chilli strains CCS's ability to build their capacity and facilitate their entry into the value chain. Existing farmer groups should be used to introduce new entrants

to the production and marketing skills needed. The farmers should be encouraged to set up similar value chains for other farm produce.

More information

***Joseph M Ndegwa**, Nakuru RegionInter- Diocesan Christian Community Services, acknkrccs@africaonline.co.ke; jmundegwa@yahoo.com; fdevelopment7@gmail.com*

***Joseph Njoroge Ndirang'u**, farmer, Laikipia West*

5 Improving chain liquidity

THIS CHAPTER TURNS TO instances where financial agents have linked with an existing chain to provide new services that build on and improve the chain flows. It includes four cases:

- **"White gold" from Tanzania** looks at a value chain for organic cotton set up by Remei AG, a Swiss company, and BioRe, its Tanzanian partner. It shows how Remei and BioRe obtained support from Triodos, a leading socially responsible bank, to provide the financing needed to expand the chain and purchase cotton from growers in northwestern Tanzania.
- **Reducing poverty by strengthening the sal leaf chain in India** describes efforts by UNIDO, a United Nations agency, to organize federations of workers who collect the leaves of certain trees and turn them into disposable plates. Organizing these workers and working with traders was key to their getting loans from local banks that were tailored to their needs, rather than inappropriate and poorly managed forms of microfinance.
- **Micro-factoring: Instant payment on delivery of tea in Kenya** shows how Biashara Factors Limited provides an innovative financing arrangement known as "invoice factoring" to small-scale tea producers. By bridging the gap between delivery and payment, Biashara provides farmers with much-needed capital and overcomes a bottleneck in the chain. This case also illustrates the use of mobile phones for payments – a service pioneered in Kenya that will surely be of huge importance worldwide.
- **Financing rice farming in Rwanda** relates how CAF Isonga, a microfinance institution, works with a growers' cooperative to provide a variety of financial services for rice farmers. The collaboration with the co-op and a payment system that allows it to deduct repayments automatically give CAF Isonga the assurance it needs that borrowers will not default on their loans.

"White gold" from Tanzania: BioRe organic cotton

"PAMBA!"

"Dhahabu nyeupe!"

"Again, all together! *Pamba!*"

"Dhahabu nyeupe!" This time everyone punches their fists in the air in unison.

A political rally? No – this is the start of another session at BioRe Tanzania Ltd.'s farmer training facility in Meatu district, northwestern Tanzania, the country's main cotton-growing region. *Pamba* means "cotton" in Swahili, and *dhahabu nyeupe* means "white gold".

There are 45 farmers in the hall. Today they will learn about quality control - an important topic for organic cotton growers. There will be a short session in the training hall, then the group will go outside to see BioRe's demonstration fields, where the buds on the cotton plants are nearly fully open, revealing the brilliant white fibre inside. The farmers will learn how to stop the seed cotton from getting contaminated at harvest.

The farmers of Meatu district were not always so enthusiastic about cotton. It is the main cash crop in the area, but when the sector was liberalized in 1993, the previous marketing system collapsed, leaving farmers with piles of unsold cotton. Many farmers still keep receipts from that time in the forlorn hope that maybe one day they will get paid for the bales of cotton they delivered to the warehouse just before it closed.

Figure 5.1 Location of Meatu, Tanzania

Establishing a new chain

Things are different now, at least for 2,000 cotton growers in Meatu district. These growers cultivate 11,000 ha of

Box 5.1 From field to wardrobe

A long and complex chain transforms the armfuls of white fluff produced by farmers into shirts and trousers. In general, these are the main steps:

1. **Farming** Farmers grow cotton, producing a raw material called "seed cotton". This contains both cotton seeds (67% of the weight) and fibre (33%).
2. **Seed cotton trading** Traders and purchase agents buy the seed cotton from farmers for bulking and transport to the ginneries.
3. **Ginning** In the ginning process the seeds are removed from the fibre (known as "cotton lint"), and the lint is pressed into bales weighing about 200 kg. Some of the seeds are retained for farmers to plant in the next season. The rest is sold to oil mills, where they are crushed into oil (used for cooking) and seedcake (used as animal feed).
4. **Cotton lint trading** The cotton bales are traded on the world market, often through forward contracts.
5. **Spinning** The cotton bales are sent to a spinning mill, where the cotton lint is processed into yarn. One kilogram of cotton lint produces approximately 650 grams of yarn.
6. **Yarn trading** Yarn is bought and sold by specialized yarn traders.
7. **Textile manufacturing** The yarn is dyed, woven, knitted, cut, stitched and printed to produce clothing and other products.
8. **Retailing** Retailers sell the resulting end-product to consumers.

cotton, producing around 8,000 tons of seed cotton a year for BioRe, Africa's leading exporter of organically certified cotton lint. That has a farm gate value of $3.5 million, or a free-on-board export value of $5 million.

This chain is coordinated by Remei AG, a Swiss firm that produces and markets organically grown textiles. The garments are sold in supermarkets and fashion outlets in Switzerland, France, Germany and other western European countries. Remei had set up BioRe to secure its supply of certified organic cotton lint. Starting as a project, BioRe is now a separate company. Remei AG buys lint from BioRe, and has a similar relationship and history with another firm in India.

Why Remei did chose to reach out and work directly with farmers, rather than relying on the cotton trading system? Cotton is one of the world's most important agricultural commodities, involving millions of farmers (Box 5.1). But the crop is associated with various problems: intensive use of pesticides, soil degradation, depletion of water resources, child labour and rural poverty. Remei wanted to do business in a different way and established a sustainable business model based on organic farming, fair treatment of farmers and textile workers, responsible entrepreneurship and genuine partnering among all actors in the cotton chain. It wanted to produce textiles that satisfy demanding consumers, preserve the environment, and enable a better quality of life for small farmers.

Remei AG set up BioRe in Tanzania as a project in 1994 with just 45 farmers. The business has grown steadily; BioRe now works with over 2,000 farmers, and has a staff of 66. It was converted from a project into a subsidiary of Remei AG in 2002, and in 2006 into an independent firm. BioRe retains very close links with Remei: the Swiss firm buys all of its output, provides financial services, and coordinates the value chain. Remei AG and BioRe want to make BioRe fully independent, establishing a truly Africa-based company with local management and, ultimately, the farmers as shareholders.

BioRe works in 15 villages within a 30 km radius of the company's head office in Mwamishali, in Meatu district. In each village BioRe has a local office with a supervisor and several extension workers, who each serve 50 farmers.

Growing the white gold: The farmers

Ninety-four percent of Meatu's households depend on rainfed agriculture for their livelihood. Each family has about 80 acres (32 ha) of land. On 10–12 acres (4–5 ha) they grow cotton, yielding $1,250/year, which is 80% of their income. On another 6–8 acres (2.4–3.2 ha) they grow sorghum, maize and legumes, mainly for home consumption. The rest of the land is left fellow for grazing. Sixty percent

Figure 5.2 Cotton is a major source of income for people in Meatu

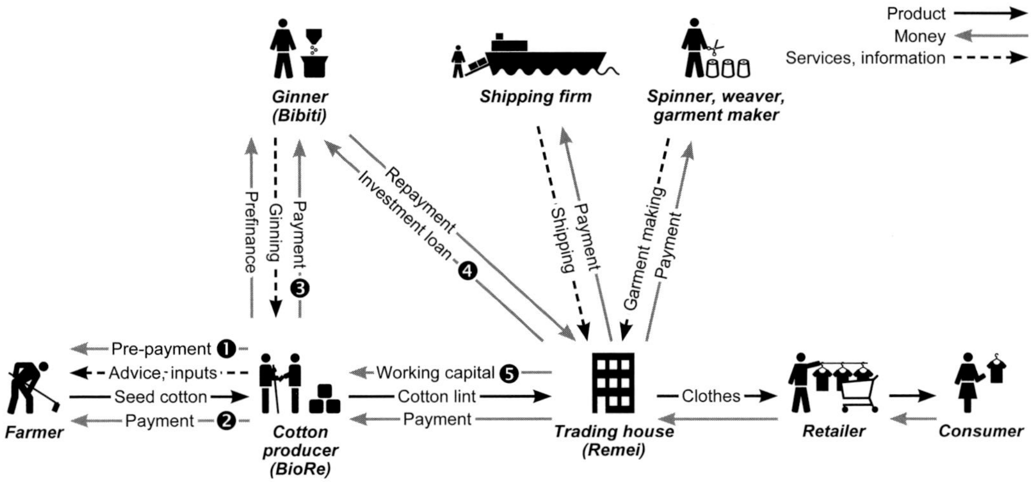

Figure 5.3 Initial BioRe/Remei cotton value chain

of Meatu's population live in poverty, and one-third of the children suffer from malnutrition.

The cotton season starts in September with the registration and contracting of farmers who want to grow cotton for BioRe. The contract obliges the farmer to use only organic production methods, follow the advice of BioRe's staff, and deliver the entire output to BioRe. The company, in turn, undertakes to purchase the entire crop, provide seeds and bio-pesticides, and offer training and technical assistance. The contract is for 5 years, but the farmer can terminate it earlier if he or she wishes. BioRe can annul the contract only if the farmer violates the contract, for example by spraying chemical pesticides.

The farmers sow cotton in the rainy reason, from November to February. They normally sow a little at a time, at intervals of a few weeks, to spread the risk of drought. BioRe's extension officers visit the farm twice a month to provide technical advice, check that the organic regulations are being followed, and estimate the expected yield. The organic certification agency also makes random inspections.

BioRe encourages farmers to work together. The company sets up farmer field schools in each location, where farmers learn about cultivation techniques, pest control, and other matters. One farmer is appointed as location leader. He or she is regularly trained on BioRe's demonstration farm, and is expected to transfer this knowledge on to the other farmers in his location. The company also supplies implements such as ox-drawn weeders for the farmer groups to use collectively.

Harvesting, ginning and export: BioRe

The harvest season is from May to July. The farmer stores the seed cotton at home waiting for the right price to sell. He or she then gathers the cotton in a cloth, usually 60–100 kg, and takes the cotton to the BioRe village office, where it is sieved to remove sand and other foreign material, inspected for dryness and quality defects, and weighed. The farmer is then paid in cash and goes back home.

Each BioRe village office can store up to 60 tons of seed cotton. A lorry takes the cotton to Bibiti Ginnery in the nearby town of Mwanhuzi. This ginnery separates the seeds from the lint, presses the lint into bales, and transports them to the port of Dar-es-Salaam. Part of the seed is kept for sowing the next season; BioRe sells the rest to mills that process it into oil and cake.

BioRe has a partnership agreement with Bibiti to provide all post-harvest services, including ginning, transport and shipping. For these services, Bibiti receives a fixed fee per ton of cotton. Bibiti works exclusively for BioRe: a convenient arrangement for BioRe as it handles no other cotton, so there is no risk of contamination, and the integrity of BioRe's organic cotton chain is guaranteed.

From spinning to clothes: Remei AG

Remei AG buys the cotton free-on-board at Dar-es-Salaam, paying BioRe by bank transfer. The cotton is shipped to spinning mills in Europe, Asia and Africa, where Remei AG contracts with specialist firms to spin it into yarn, weave it into cloth, and produce garments that Remei AG co-designs with large retailers and fashion brands in Europe.

Chain liquidity

The actors in the cotton chain rely heavily on BioRe to finance their business operations. The farmers have very little capital. Cotton is their only cash crop, so by the time it is harvested they urgently need money. If BioRe did not pay in cash, the farmers might sell their crop to other buyers – even at a lower price. In 2008, BioRe needed $3.5 million in cash to pay its contract farmers.

Prefinancing the crop BioRe has an arrangement to pre-finance the crop. For each kilogram of seed cotton a farmer delivers, BioRe pays an input subsidy into a special bank account. The farmer receives a passbook showing how much is in this account. The next season, the farmer can use this passbook to buy seeds and bio-pesticides from the company (❶ in Figure 5.3). Such a passbook scheme used to be common throughout the cotton sector in Tanzania, but it fell out of use; since 2008 BioRe has been the only company still offering it.

For new farmers who do not yet have a passbook, BioRe is willing to pre-finance the crop through a loan at zero interest rate. But this is risky for BioRe: the farmers have no experience in organic farming, and it is hard to assess how trustworthy they are. So many drop out of BioRe's scheme. In 2007, 23% of new farmers did not deliver their cotton to the company, even though they had received the loans – a loss of $7,000 to BioRe. Despite these losses BioRe continues to pre-finance farmers who are converting to organic production, otherwise it would be nearly impossible to attract new farmers.

Paying farmers Distributing so much cash to thousands of scattered farm households is major operation. Every day in the buying season, BioRe withdraws the equivalent of $50,000 in cash from the National Microfinance Bank, the only bank in the district. As BioRe needs small bills, these are two large bags of notes. The money is brought under police escort to BioRe's office, where it is divided among the 15 village supervisors. The supervisors take the cash on their motorbikes to their villages to buy the cotton from the farmers. At night they come back to the office to hand in the purchase slips certifying the payment to the farmers. At the same time they tell the office how much money they will need the next day. Each day the same operation is repeated (❷ in Figure 5.3).

The costs of handling the money are high, and the risks of robbery are significant. Also, there is the risk that supervisors and farmers collude in signing payment slips which are not backed by an actual delivery of seed cotton. In the cotton industry normally around 1% of the cotton gets "lost" in this way. But BioRe has no alternative ways to pay the farmers. There are no financial agencies in the villages where the farmers live, so they have no accounts to which money can be transferred.

Working capital for the ginnery Like the farmers, the ginnery is also short of cash. Every year BioRe lends it money to buy spare parts and maintain the factory. In February Bibiti presents a shopping list of spare parts, which BioRe pre-finances to a maximum of $60,000. Then in the ginning season BioRe continues to make down-payments, twice per month. The down-payment is disbursed after the ginnery reports how much cotton lint it has produced – so BioRe can avoid overpaying for the services the ginnery has provided. The ginnery submits the invoice for its services only at the very end of the season to keep as much cash as possible in the value chain when it is needed (❸ in Figure 5.3).

Capital investment for the ginnery Some years ago, Remei gave the ginnery an investment loan of $400,000 to buy and install new equipment, to be paid back over 4 years (❹). Part of the deal was that the ginnery would give BioRe preferential treatment in processing its seed cotton. In fact, the ginnery stopped serving other clients, so now processes only BioRe's cotton.

Working capital Remei provided all the working capital that BioRe, the ginnery and the farmers needed. Every week Remei transferred a significant amount of money to BioRe's bank account (❺), so that BioRe could pay its staff and bills, and pre-finance the farmers and the ginnery.

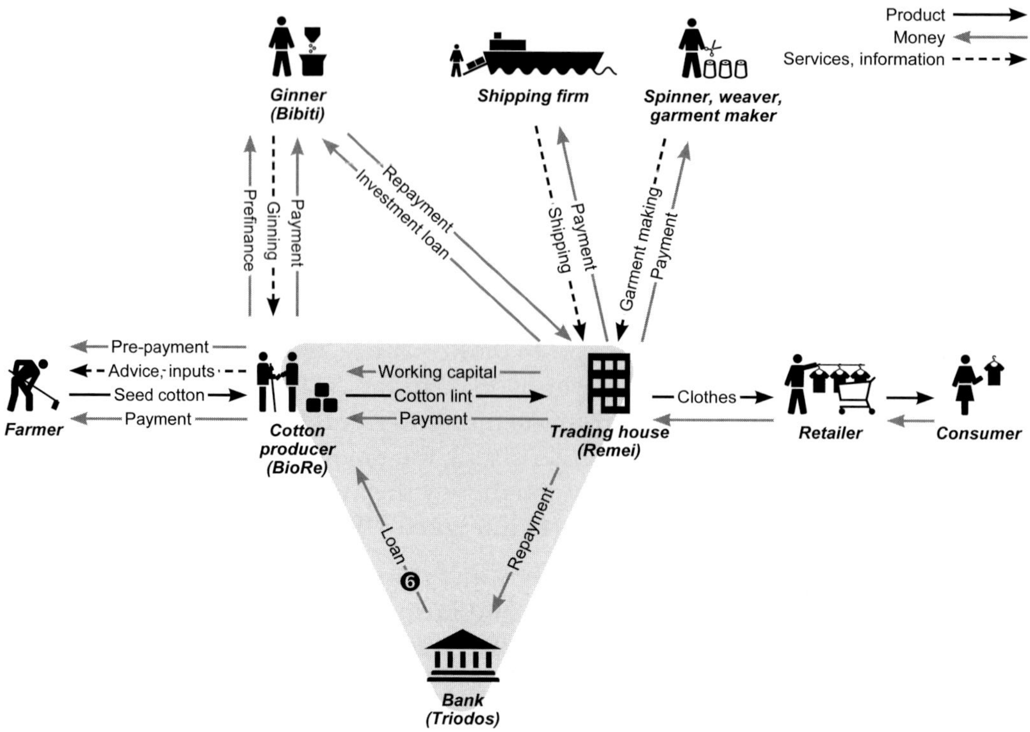

Figure 5.4 The current BioRe/Remei cotton value chain

A new source of finance

This arrangement was far from ideal. For Remei AG it was a heavy financial burden, even though the company is financially solid and has a respectable turnover. Remei AG financed its own operations, all cotton and textiles in stock in the chain, and the operations of BioRe as well as Remei AG's Indian supplier. The more Remei AG grew, the more difficult it was to get enough credit from banks to finance these worldwide operations. In addition, Remei AG's vision is for the BioRe companies to become fully independent. They should not depend on Remei AG for financing. So Remei AG has been looking for new sources of finance for the BioRe operations in Tanzania.

In 2004 Remei AG asked the Triodos Sustainable Trade Fund, part of Triodos Bank in the Netherlands, to pre-finance the cotton operations in Tanzania. Triodos Bank is one of Europe's leading socially responsible banks. Its mission is to finance companies and projects that benefit people and the environment. The Triodos fund assists certified organic and fair trade producers in pre-financing their export contracts with buyers in Europe and the United States.

Triodos shared Remei AG's vision that BioRe should have the opportunity to become independent, so it decided to help the Tanzanian firm build direct relations with financers. The relationship went through a series of stages.

Table 5.1 Details of Triodos loan to BioRe/Remei AG

Product and financial flows	
Purpose	Pre-finance of export contract between BioRe and Remei AG; any other use is considered a default
Amount	$2,325,000
Period	10 months
Disbursement	Disbursed directly to BioRe Tanzania in three instalments on 15 Jul (22%), 1 Aug (43%) and 1 Sep (35%)
Repayment	Repaid by Remei AG in four instalments on 15 Nov (4%), 15 Jan (19%), 16 Feb (39%) and before 1 Apr (38%)
Interest rate	8% annual
Transaction costs	1% administration fee
Risk management	
Securitization	No hard collateral Securitization based on sales contract between BioRe and Remei AG Loan is maximum 60% of the free-on-board value of produce
Liability	BioRe Tanzania and Remei AG are jointly and severally liable
Information flows	
Information required to apply	Most recent audited financial statements Most recent interim financial statements Cash flow prognosis for new season and projected profit and loss account Copy of the export contract and certificate of organic quality Review of previous season: realized versus expected yield, number of contract farmers and hectares; problems and claims regarding produce quality; development of market prices; competition in sourcing the produce; realized investments; financing from others; changes in ownership and management; experiences in transport and shipping Outlook for coming season (same issues as above)
Information required during season	Monthly project monitoring report specifying seed cotton and lint cotton in stock, seed cotton readily available for sourcing, and cotton lint shipped abroad
Time lag between application and payment	1 month

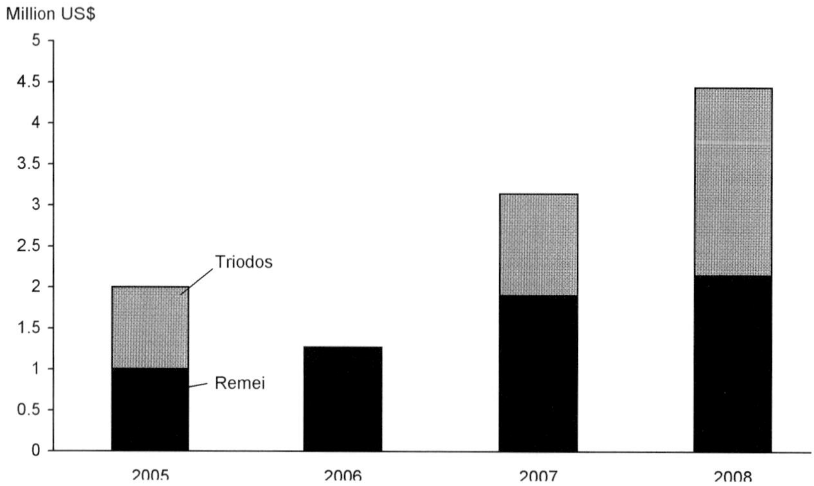

Figure 5.5 Trade finance loans to BioRe Tanzania

- **2005** Triodos provided Remei AG with a loan of €800,000 (about $1 million). Remei AG transferred the money to BioRe, and after 7 months Remei AG repaid the loan to Triodos.
- **2006** Drought severely affected the crop, so BioRe and Remei AG decided not to apply for a loan.
- **2007** Triodos provided a new loan, this time for €900,000 (about $1.2 million), again indirectly to BioRe through Remei AG.
- **2008** The partners had built trust and decided to go a further step. The loan was converted into US dollars, and $2,325,000 was disbursed directly to BioRe (❻ in Figure 5.4). Both BioRe and Remei AG are jointly liable for the loan, and Remei AG will repay it. The loan period has been extended to 10 months, and the dates of disbursement and repayment have been adapted to better accommodate the trading season. Table 5.1 shows details of this loan.

Meanwhile, Remei AG has also continued to finance BioRe (Figure 5.5). The additional finance from Triodos has made it possible for BioRe to expand its operations in Tanzania and increase the number of farmers it buys cotton from. Overall loans rose from $2 million in 2005 to $4.5 million in 2008, with half coming from Triodos in both years.

The partnership with Triodos has transformed the financial flows in the cotton chain from chain liquidity to a value chain finance triangle (Figure 5.4). Triodos's willingness to finance the chain is based on the strong relationship between Remei AG and BioRe, which is in turn based on proven relationships further up the chain (between BioRe, Bibiti and the farmers of Meatu district) and down (the links between Remei AG, the firms that spin the yarn and turn it into clothes, and the retailers). These relationships, plus strong market demand for organic cotton, mean that Triodos can be confident that the value chain will continue to generate revenue.

Mitigating risks of financing cotton

Exporting cotton from Tanzania is a risky business in the eyes of many banks. They see various factors that may affect an exporter's capacity to repay a bank loan (Table 5.2):

- **Climate.** Cotton farming in Meatu district is rainfed, and yields depend heavily on the rainfall. Every few seasons there is a serious drought. When yields are low, the exporter may not have sufficient revenue to repay the loan.

Table 5.2 Risk analysis for the organic cotton chain in Tanzania

	Source of risk	Risk level	Mitigation strategy
Production	Drought	●●○	Accurate yield forecasts, based on visits to farmers every 2 weeks
	Contamination of chemicals ("drift")	●○○	Grow hedges or leave gaps near non-organic cotton fields Compliance checked by inspectors
Price	Price fluctuations	●●○	Purchase guarantee from Remei AG with continuous price adjustments based on "open-book budgeting": Remei AG looks at BioRe's production costs and adds a small profit margin to calculate sales price
Market	Lack of market demand	●●○	BioRe has purchase guarantee from Remei AG BioRe commits to buying only 80% of actual yield of farmer
	Market may not pay premium for cotton not yet fully certified as organic	●○○	BioRe will continue to buy such cotton and pay the input subsidy, but will stop paying the organic premium
Default	Side-selling to competing traders	●○○	Offer premium price to farmers Close relationship with farmers based on transparency Short storage period (weekly) Budgeting accounts for 9% default
	Non-compliance with organic regulations	●○○	Visits to farmers every 2 weeks Internal and external inspection Continuous training
Currency	Exchange rate US$–TSh	●●●	Loan and payments are in dollars, so currency risk reduced to 2 months that cotton is in BioRe's hands

Risk levels: ●●● High ●●○ Medium ●○○ Low

- **Price fluctuations** The price of cotton is set by the world market. In previous years prices had risen steadily, but in 2008 they suddenly dropped. If BioRe has a lot of unsold lint in stock when the price falls, this implies a huge loss.
- **Default** Farmers may sell their cotton to competing traders, even if they have signed a contract. They may not meet organic requirements or quality standards. Default by farmers may lead to a shortage of cotton and revenues for the exporter.
- **Currency risks** BioRe buys seed cotton in Tanzanian shillings and sells lint in US dollars, so is at risk of currency fluctuations.

These factors meant that many banks are reluctant to provide loans to cotton exporters such as BioRe. BioRe requested loans from several Tanzanian banks, but they required mounds of documents, safeguards such as the firm's office and training centre as hard collateral, as well as personal guarantees from BioRe's and Remei AG's managers.

Triodos does not demand such assurances. Its loan to BioRe is not secured by hard collateral or by personal or corporate guarantees. The only document underpinning the loan is the sales contract between BioRe and Remei AG, specifying the amount and price of lint to be sold. Triodos' decision to extend the loan is not based on securities, but on the trust that the cotton lint will be produced and sold, and will generate sufficient profit to repay the loan. In financial jargon, the loan is based on cash flow projections, rather than securities from assets.

There are three key strengths in the BioRe/Remei AG chain:

- **BioRe's management capacity** Over the years BioRe has shown it can deal effectively with the problems and risks of cotton production and export. Field officers pay monthly visits to the farms, so the company has realistic forecasts of expected yields. BioRe has close relationships with the farmers, offering 5-year contracts, inputs, training, and technical assistance. And BioRe builds in an assumed default rate of 12% into its yearly plans in case of possible setbacks.
- **Partnership between BioRe and Remei AG** The companies have a sales contract with fixed prices and volumes. So the risk of market fluctuations is eliminated for BioRe and the farmers. Beyond the contract, there is strong commitment and interdependence between the two companies. Any issue will be resolved in close partnership. Remei AG is a financially solid company with a strong balance sheet.
- **The integrated textile chain** All companies in the BioRe chain are partners who have been working together for many years. They are committed to each other and cooperate smoothly to solve problems. As chain manager, Remei AG ensures compliance with quality standards, open communication and efficient coordination in the chain.

These strengths of the BioRe cotton chain effectively mitigate the risks of cotton trading. Many banks fail to recognize this. But Triodos specializes in financing certified export chains. Its staff can recognize the strengths of the BioRe cotton chain, and weigh these against the risks. That is why Triodos is willing to pre-finance the cotton chain without demanding hard securities, as local banks would do.

Setting prices

In Tanzania, the farm gate price for seed cotton is set in the following way. Before the harvest starts, the floor price of the season is determined by the cotton traders in a meeting of the Tanzania Cotton Board. The floor price depends on various factors: the world market price for cotton, the costs of oil and other inputs, and the exchange rate of the Tanzania shilling against the US dollar. Once the floor price is set, the free market is left to do its work. The traders compete with each other in buying the cotton from the farmers, so the prices normally go up during the season. For that reason, farmers only sell little by little, enough to have sufficient cash money to pay their urgent bills, while waiting for the highest price possible.

BioRe's purchase price is an average of the prices offered by competing traders in the villages where the company operates. Seed-cotton traders are obliged by law to publicly announce their buying price on a billboard outside their office. BioRe gathers this price information every day to calculate its own purchase price. On top of this, the company offers a fixed price premium of 15% on the average market prices of the last 5 years. The average traders' purchase prices have risen from around TSh 160 per kilogram in 2000, to about TSh 480 in 2008. BioRe's prices have risen from TSh 180 to about TSh 550 in the same period.

BioRe's and Remei AG's strategy is to pay the farmers as much as possible, at the lowest cost possible. The organic premium and the input subsidy for the farmers

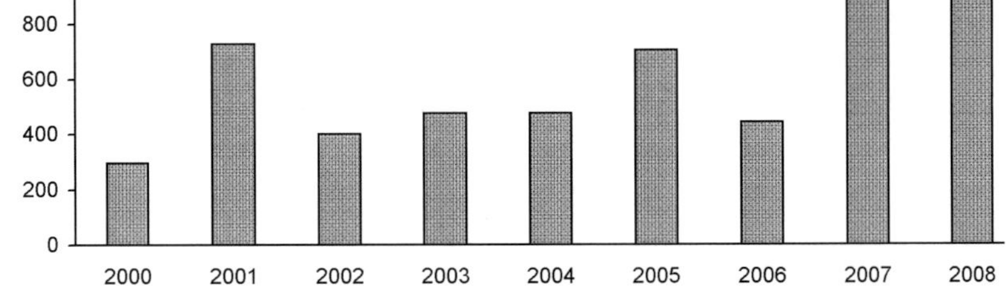

Figure 5.6 Average income of BioRe's farmers

Table 5.3 Income and expenses of organic and conventional cotton farmers

US$ per hectare

	Organic (BioRe) farmer	Conventional farmer
Income		
Sales of seed cotton	258.12	258.12
Organic premium	32.40	–
Total income	290.52	258.12
Expenses		
Seeds	5.00	5.00
Land preparation	25.00	25.00
Weeding	75.00	75.00
Harvesting	32.40	32.40
Inputs	3.13	8.00
Total cost	140.53	145.40
Input subsidy	8.10	-
Net expenses	132.43	145.40
Net earnings	**158.10**	**112.72**

are not paid for by BioRe, but by Remei AG from its annual profits. So the extra payments to BioRe's farmers are not added to the cost price of the cotton. Why is this smart? Because usually chain actors automatically add a percentage mark-up to the value of the product, and taxes are also calculated on the product's value at each stage. By "lifting out" these costs and paying them at the end, Remei AG avoids inflating the price of the product at each stage in the chain. That keeps the price of the final product low, making consumers more willing to buy it.

Benefits

Farmers BioRe's farmers not only receive a price premium; as organic producers they also avoid the costs of chemicals, while their yields are similar to those of conventional cotton farmers. That gives them an income 40% higher than conventional farmers (Table 5.3). In addition, thanks to the training and advice from BioRe, farmers have learned to produce more efficiently, and they are now able to produce larger areas of cotton. As a result, their income has boomed (Figure 5.6).

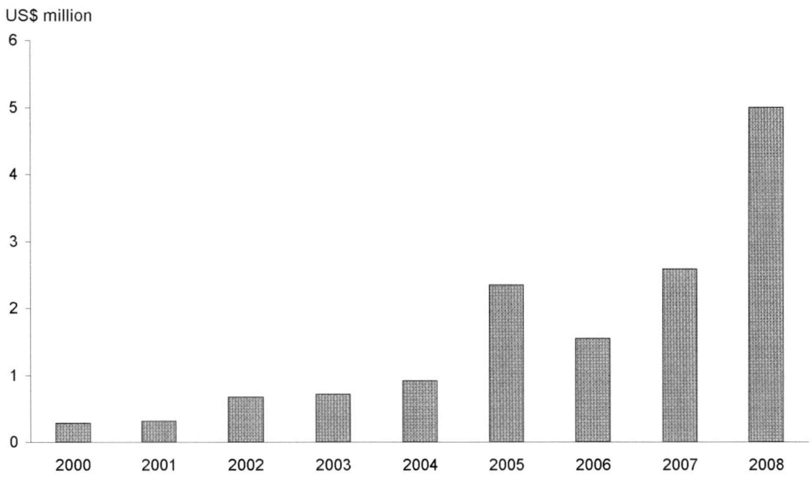

Figure 5.7 Annual turnover of BioRe

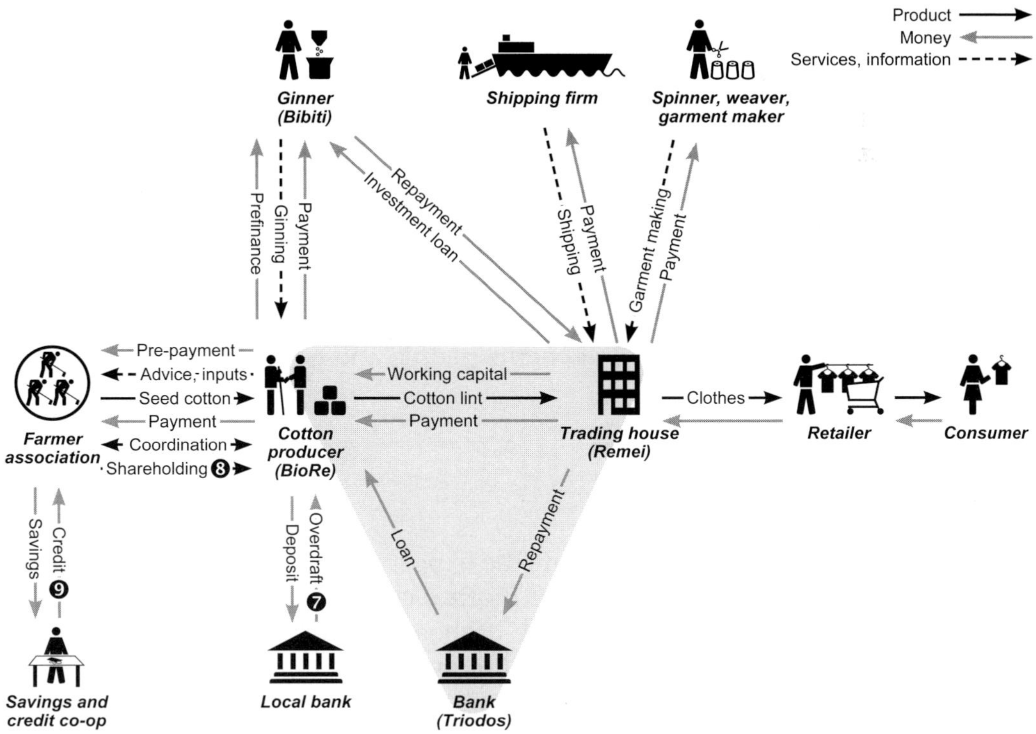

Figure 5.8 The future of the BioRe/Remei AG cotton value chain

BioRe's farmers also report other benefits. They obtain higher yields in other crops. Their soil is in a better condition since they started rotating cotton with other crops. The soil has more organic matter, which increases the soil capacity to retain humidity. So a period without rainfall is not as fatal for them as it is for other farmers in the district. Finally the farmers save a lot of time for their family, as they learned to use oxen for ploughing and weeding.

Other local people BioRe now employs 66 staff, most from local villages. It is the only employer in Meatu district which provides such attractive salaries and education opportunities. BioRe's training centre is developing as a regional centre of competence. In times of need the company supports the local community. During the drought of 2006 the company provided daily lunches to 7,000 schoolchildren for 3 months.

BioRe BioRe has enjoyed a spectacular increase in turnover (Figure 5.7). It is a key part of a chain that maintains high social, environmental and quality standards. The chain has grown markedly, outperforming conventional cotton chains, and generates profits for all actors.

Bibiti The Bibiti ginnery's contract with BioRe, along with the prefinancing and loans it receives, enables it to run efficiently and profitably. Without this relationship, its ginnery would most likely stand idle.

Remei AG Remei AG and its downstream partners enjoy significant growth and sound profit margins. In the last decade Remei AG has switched fully to organic cotton. Its turnover has remained more or less stable at $25 million, but its profitability has more than tripled.

Consumers Consumers are able to buy high-quality garments that fulfil highest standards of sustainability at a price which is not so much higher than conventional clothing.

Triodos Bank Triodos has a good client in BioRe. There is a smooth working relationship; so far there have been no problems in reporting, communication or repayment, so the bank does not need to spend much time or effort in managing the loan. Most important, the loan is fully aligned with the bank's mission – it is a financial product that directly benefits people and environment, in a region of great need.

The future

BioRe and Remei AG jointly determine the export price of BioRe's lint by "open book" calculation: the partners look at BioRe's costs of production, then add a small profit margin. In 2008 BioRe's free-on-board export price was $1.73/kg of lint. This was significantly higher than the average Tanzanian export price of $1.58. In other words BioRe is not yet competitive: it would not survive if Remei AG were not willing to pay a higher price. The main reason of BioRe's higher costs is its investment in training and technical assistance to the farmers. BioRe needs to find a way to reduce its costs and produce cotton at a fully competitive price. To accomplish this, BioRe is pursuing a strategy with the following elements (Figure 5.8).

Involve local banks more closely The loan from Triodos is very attractive. But it would still be good to develop a stronger relationship with local banks. One important thing for BioRe is to get an overdraft account at a local bank, where BioRe can draw small amounts of short-term credit, for example, when a client pays late (❼ in Figure 5.8). Another issue is to find a better way to pay the farmers. Rather than paying in cash, BioRe would prefer to transfer money to farmers' accounts.

Establish farmer associations BioRe wants to organize the farmers into groups that can help to distribute seeds and inputs, collect and inspect the cotton, and jointly deliver to BioRe. The groups can also take care of training activities. In the long term, the groups can become shareholder in BioRe. In Remei AG's Indian partner, this is already happening (❽ in Figure 5.8).

Establish savings and credit cooperatives By establishing savings and credit co-ops, the farmers could provide some of their own financial services. For example, they could save a small amount each week, then loan this money to members. The farmers could invest their loans in cotton production, or use it for other purposes (❾ in Figure 5.8).

BioRe has learnt that it takes time to build a healthy enterprise. After 14 years the company is still not in a stable situation. To achieve high standards, the company has invested heavily in training of farmers and staff and built an internal inspection system. This created a heavy overhead. The overhead now needs to be reduced by empowering the farmers and ensuring they take on more responsibilities.

Further information

__Niranjan Pattni__, Managing Director, BioRe Tanzania Ltd., niranjan.pattni@biore-tanzania.com

www.remei.ch

www.coop.ch/naturaline

www.triodos.com

Reducing poverty by strengthening the sal leaf chain in India

WEDDINGS IN INDIA TEND to be massive undertakings. Hundreds of well-wishers come to present gifts to the bride and groom. Naturally, all the guests must be fed. Behind the scenes, a small army of cooks prepares platter upon platter of food. But how to serve it? Few households or hotels have enough crockery to serve so many people at one time.

The answer is to use disposable plates and cups. Unlike their plastic or coated paper equivalents in the West, in India these are often made of leaves. They are cheap, disposable and biodegradable. Apart from weddings, they are also used in festivals and ceremonies, and to make offerings to Hindu gods.

These cups and plates are made from the leaves of the sal tree (*Shorea robusta*) – a tall timber tree common in eastern and central India. The oval leaves are 10–25 cm long and 5–15 cm wide – too small to hold more than a morsel of food, so they must be sewn together and then pressed into shape to make a rigid plate or cup.

Used by some of India's wealthiest people, these items are made by some of its poorest: tribal women in the states of Orissa, West Bengal and Jarkhand (Figure 5.9). Collecting leaves and making plates and cups is a major source of income for thousands of people who live in this region's forests, as well as for hundreds of traders. Many tribal families depend on collecting and selling sal leaves for over half their income. There are few figures on how many leaves they collect and sell each year. At a rough estimate, the total wholesale value of sal leaf plates and cups in Orissa alone is about $200 million.

Figure 5.9 Location of the sal leaf industry in India

In the district of Mayurbhanj, many sal leaf collectors, processors and traders cluster in and around the district capital of Baripada. In 2006, UNIDO, a branch of the United Nations that works on industrial development, began working

Figure 5.10 Stages in sal leaf production

in this district as part of its Cluster Development Programme. The aim of this programme was to reduce poverty by strengthening the sal leaf value chain. The programme reached around 10,000 households living in 25 villages in a forest area of 10,000 hectares.

The sal leaf value chain

Figures 5.10 and 5.11 show the main stages and actors in the sal leaf chain.

Collectors Most of these are tribal women. They collect sal leaves in the forest for about 8 months of the year, dry them and stitch them together into *dwipatris* (twin leaves) or *khalis* (raw plates). These stitched leaves become the raw material for making sal leaf plates and cups. They bundle them and sell them to local agents, earning about $0.24 to $0.30 per day. This is barely enough to survive on, and even this income is seasonal. During the low season from September to January when leaves are scarce, life gets even tougher. The collectors have no savings, so they look for farm work, collect edible tubers from the forest, or look for other sources of income.

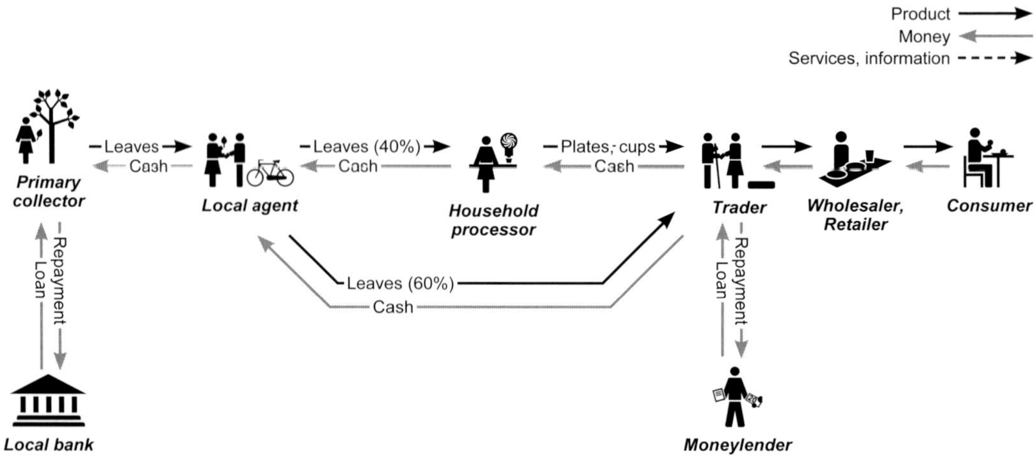

Figure 5.11 Traditional sal leaf chain

Local agents Agents, also called "cycle wallahs", are men who go door-to-door buying the stitched leaves from collectors, paying them in cash. They sell bundles of leaves on the local market or directly to processors who convert them into cups and plates. The agents are not tribal people, and most live in nearby towns. They are better off than the collectors, earning US$90–100 per month. They trade in leaves part-time; many also run small businesses such as grocery shops as additional sources of income.

Processors Processors are women who work at home to make cups and plates from the leaves. Sometimes their teenage daughters stay home to help out. They operate one or two electric pressing machines. They buy stitched leaves from the local agents or local markets, press them and sell them to traders. They have minimal working capital: just enough to buy raw leaves. They earn $20–30 a month if they use family labour. For most, this is their main source of income. Most live around the market town of Betnoti, one of India's biggest centres for the trade in sal cups and plates.

Traders Traders are men who buy the cups and plates from the processors. Some also process leaves themselves. They hold a license to operate as a sal trader and a transit permit to transport the finished product to wholesalers and retailers. Larger traders have direct access to outside markets and own offices and warehouses to store the finished product. The average annual turnover of a processor–trader is $20,000 to $30,000.

Problems in the chain

Collectors

Inappropriate credit Pushed to fulfil government credit-disbursement targets, local banks often offer loans that are too large for borrowers to manage. This system leads to fraud and massive defaults, especially in areas such as Mayurbhanj, where the indigenous tribal people have little education and are pressed to take on loans of $1,000 or more. Sometimes they are not even aware that they have become borrowers: they regard the loan as a gift.

In the 1990s, village health workers encouraged women to form groups of up to 20 members and to start saving regularly. The women opened group accounts with a local bank and lent small amounts from their group's savings to one another. After a group has done this successfully for about 6 months, the banks may offer it a loan, giving it a bigger pool of money to lend out to its members. By 2009, there were about 3 million such groups in India, with around 45 million women members; many groups had accumulated funds of several thousand dollars. But unfortunately this programme has also become target-driven. The banks have pushed credit loans onto the self-help groups to meet their loan targets, leading borrowers into a debt trap.

Inappropriate use of loans The collectors often use their loans poorly. These loans are meant for investing in sal leaves, *sabai* grass (used to make rope), poultry, livestock rearing or other enterprises, but borrowers lack the skills or materials to do this successfully. Instead of investing it in production, many spend it on consumption, sometimes using large amounts for food and clothing during traditional celebrations, or on alcohol.

Lack of business skills In addition, the collectors lack skills such as book-keeping, accounting and business planning. Unlike some financial institutes which provide "hand-holding" services with a loan, these government-funded loan schemes give lenders no extra services. As a result, most collectors default.

The collectors often do not grasp what a default is, nor do they feel the consequences. They have no assets that could be repossessed. For their part, the bankers merely erase the defaults from their books by giving the collectors a new, larger loan to pay off the first debts and give them a little extra money.

Reduced availability of sal leaves. In most villages, collectors cannot collect enough leaves because of illegal tree felling and unsustainable harvesting practices.

Lack of market information. Collectors lack access to markets or market information. Local agents take advantage of this by paying them below-market rates and not telling them about market conditions.

Processors

No access to credit Processors have the opposite problem to collectors: getting enough working capital to build their businesses. Banks are not interested in financing them because they have no land, no collateral and no documentation such as financial statements on which to base a loan. They are not considered a target group in government policies, so they do not have access to credits from the government banks.

Low productivity due to outdated technology Without working capital, processors must continue using old pressing machines. The machines are dangerous: processors complain of chronic leg pain and electric shocks. They are also energy-inefficient: they waste a lot of heat. Changes in voltage and temperature affect the quality of the pressed cups and plates. The machines frequently break down, causing days of lost work as processors wait for them to be fixed.

Reduced income in the low season Prices for raw materials (stitched sal leaves) fluctuate widely. During the low season when sal leaf is scarce, the price of stitched leaves may rise by 30–35%, even as the quality deteriorates. For processors, this is a difficult time. They may operate at a loss – just covering the cost of raw materials but not their labour – in order to remain connected to traders. Sometimes they are even forced to stop production. Big traders with large stores hoard stitched leaves during the high season for use at other times of the year. With some working capital, smaller-scale processors would be able to do the same, keeping their profit margins more stable.

Traders

No access to credit Like the processors, traders also lack access to bank loans. They tend to rely heavily on informal credit, such as loans from larger traders and moneylenders, which charge high rates of interest. Without working capital they cannot expand or improve their businesses, which has impacts all the way down the chain.

Competition from substitute products. Traders have failed to innovate: they do not look for ways to improve their products or build their businesses. Their earnings are starting to suffer as leaf plates face growing competition from cheaper plastic and paper plates. At the same time, a growing market for eco-friendly products presents new opportunities for products made from biodegradable leaves.

Innovating to improve the chain

Beginning in January 2006, the UNIDO programme worked with around 9,000 collectors, 200 processors and 30 traders. The programme did not deal with the local agents, who had incomes from other sources which made them less vulnerable. The programme introduced various financial, social and environmental innovations at the different stages in the value chain.

Financial innovations

Primary collectors

Training on credit and business management UNIDO counselled the self-help groups on good financial practices, including savings, internal lending (where members lend to each other through the group) and financial management (Box 5.2).

UNIDO also trained its own field staff (who are responsible for developing clusters of micro-enterprises) and the village health workers to coach the self-help groups: providing support services such as training and capacity building on bookkeeping, accounting and preparing business plans to ensure they would invest their savings and credit in business activities rather than consumption. Such hand-holding continued for one year, after which the self-help groups started regular saving and internal lending activities.

The programme organized a drama called *Das Kati*, where actors demonstrated to the villagers the consequences of taking too much credit and defaulting on loans. Three hundred collectors were taken to visit and learn from successful self-help groups outside the district.

Box 5.2 How the Dumapada self-help group took its unused loan back to the bank

In 2007, a self-help group from the village of Dumapada applied for a new bank loan. In November that year it received $600 from the regional rural bank Baitarni Grameen and distributed the money equally among its members. UNIDO organized a meeting with the group to see how they would use the loan. When UNIDO asked the borrowers what the purpose of the loan was, most did not know. Some replied that since they had no business plan, they would keep it in their savings account.

That is when UNIDO decided to provide financial counselling to help the members develop a business plan and direct the loan towards production. A half-day training on business plans and financial management was organized in which member learned to calculate their loan needs, the best use of the loan and a repayment plan. After the training, the members realized that the loan they had received from the bank was more than they needed, and more than they could use. They decided to divide $400 among the members and return the rest to the bank.

All the members invested their loan money in their sal leaf businesses and have decided on a repayment schedule of 10 months. Five months later, they were repaying their instalments regularly and earning more from their businesses.

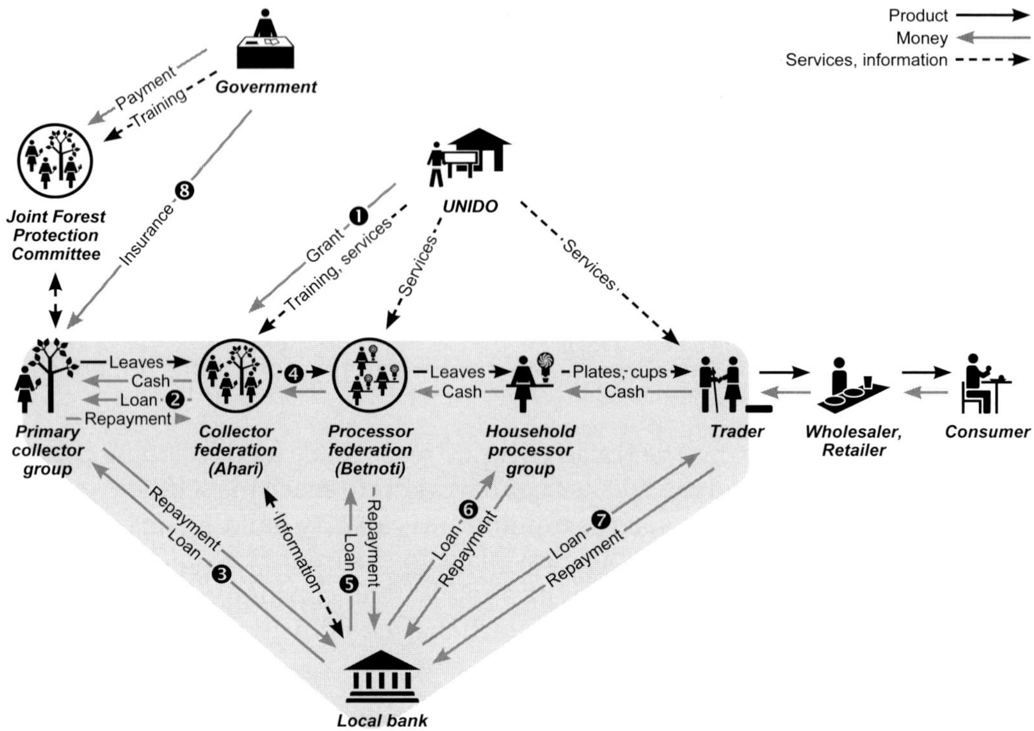

Figure 5.12 The sal leaf value chain after UNIDO's intervention

Federation loans The Ahari Federation, composed of six self-help groups, was formed to improve the management of loans among the collectors. Individual self-help groups contributed about $40 to a common fund, demonstrating their commitment to joint action. The Federation used the resulting total of $240 to get a matching grant from UNIDO (❶ in Figure 5.12).

The Federation uses this capital to give small, short-term loans to individual self-help groups (❷ in Figure 5.12) to enable collectors to stockpile leaves and wait for market prices to rise before selling them. Before, they had no choice but to sell leaves to cover their daily expenses, even though prices were low. Now the women can increase their profits and slowly build up their income and savings.

The groups receive credit based on the business plans they developed through UNIDO's training. The loans meet the actual credit needs – which are low – and the time period needed – which is short. The Federation charges its members 2% interest on loans for up to 7 days, 3% for 8–15 days; and 5% for 16–30 days. Beyond 30 days, the Federation imposes a fine. Although these interest rates are high, they are set by the groups themselves, and the profits go to the Federation rather than to a bank. The groups assist in this loan scheme by supporting their members' loan requests and pressuring members to repay on time. See Table 5.4 for details of this loan scheme.

From the start this loan scheme has worked well. Monthly repayments to the Federation were made on schedule, and the scheme showed the banks that it is possible to lend to the collectors at a reasonable level of risk.

Bank loans UNIDO sensitized the banks about collectors' credit needs and how the chain works. Within a year of the start of the UNIDO programme, and after seeing the success of the Federation loans, the banks realized that collectors need small amounts of credit for short periods - around $10–20 for up to 6 months. Based on these results, the banks started financing the collectors via the self-help groups (❸). These were production loans, intended for collectors to develop their sal businesses and to develop an alternative livelihood during the low season for sal leaf. *Sabai*, a type of grass used to make rope, is another forest product found in the region which could serve this purpose. However, collectors needed capital to invest in processing *sabai*. As with the Ahari Federation's loans, the credit amounts and repayment schedule were designed according to the self-help groups' business plans.

Access to market information The collectors used to rely entirely on the local agents to sell their products to processors. They had no access to market or to information, so the agents could get away with paying below-market rates. UNIDO took collectors to markets to learn how they work and how to gather information on a regular basis, such as the current prices for sal leaf. The Ahari Federation also started an information and communication centre at its offices where collectors can get such information. They can also call traders and processors directly to organize the sale of their product.

Integrating the value chain The self-help groups applied for a licence to trade in sal leaves. Licensing is required by the government because of over-exploitation of the forest. The collectors bring their leaves to the Ahari Federation, which sells them to a federation of processors, which in turn sells them to the processors (❹). This cuts out the local agents, so increasing the profit margins of collectors and processors.

The collectors used to earn about Rs 12 ($0.24) per bundle (the amount a collector could produce in a day) by selling it to an agent, who would sell the same bundle to a processor for Rs 23–25. Under the new system, the federation of processors pays the collectors' federation Rs 19 for a bundle; the collectors' federation keeps Rs 2 profit and gives the collectors group Rs 17; the collectors' group keeps another Rs 2 and gives Rs 15 to the collectors. The federations now act as friendly intermediaries; they reinvest this in the federation. The processors pay Rs 3–5 less per bundle, but the collectors earn Rs 3 more.

Emergency loans The Ahari Federation started offering emergency loans to cover sudden expenses such as illness. The Federation keeps cash at its offices so it can provide such loans immediately. The borrower has 10 days to repay the loan interest-free; after that period, interest is charged. The interest rate is the same as that for other loans from the Federation to its members.

Processors

To tackle the different problems of the processors, a federation of self-help groups of household processors was established near Betnoti. The Betnoti Federation members started saving money on a monthly basis and internal lending, much like the Ahari Federation of collectors.

Financing production facilities Using their savings and with support from a local organization called Gram Swaraj, the Betnoti Federation built a storage room for sal leaves, a meeting room and a common work area for processing the leaves. The total investment was $3,000, of which $2,000 was a grant from Gram Swaraj. In addition, the Federation used a grant of $1,100 from the Foundation for MSME Clusters (an NGO promoting small and medium enterprises) to buy a pressing machine. This can be used by four processors at the same time, and can generate a profit of $5–10 per day. The Federation facilities are used as a meeting place, for demonstrating new processing technology, and for training processors how to improve their productivity, reduce their costs, conserve energy and develop new products from the sal leaves.

Bank loans UNIDO organized workshops with bankers to explain how the value chain works, the opportunities and strengths of the chain. The Betnoti Federation then approached a bank and got a matching loan against the Federation's savings (❺ in Figure 5.12). The loan was used to buy sal leaves from the collectors' Ahari Federation during the high season. The leaves were held in the Betnoti Federation's storage room and then sold during the low season to processors at competitive rates.

In an effort to secure investment loans for new technology, the banks were invited to look at the strength of the leaf pressing business. The banks also wanted market information and asked the traders whether this technology would contribute to a better product. Convinced that it was a sound investment, the banks began lending money to the processors to upgrade their equipment (❻ in Figure 5.12). One major improvement was a thermostatic cut-out to prevent overheating, reduce electricity consumption and improve safety. While the traditional system consumed 2.2 units of electricity every 3 hours, the improved machine consumed just 1.2 units – a 40% reduction in energy use. Other improvements included a better cutting blade and a pressing lever that greatly reduced the physical effort of processors.

New product development In an effort to reduce leaf consumption and over-exploitation of the forest, UNIDO started motivating processors and traders to develop new products. A trader came up with the idea of using glossy paper for the bottom layer of sal leaf cups, rather than using two layers of sal leaves. For processors there were a number of advantages to this - transport costs were lower, and paper was easier to procure than leaves. Consumers also liked the hybrid cups for their strength and water-resistance, and they now fetch a better price on the market than traditional sal cups. In order to increase the scale of production, processors were trained in paper plate processing according to the specifications provided by the trader. The Betnoti Federation is working towards bringing all the processors into this initiative.

Traders

Loans One of the main reasons that banks refused to lend to traders was their lack of financial records such as balance sheets and income statements. Initially the traders were sceptical and did not want to disclose their business turnover. But UNIDO arranged meetings with bankers, took the traders on an exposure visits, and helped the traders prepare basic financial data so the bankers could appraise their businesses and lend them money. This gave banks the confidence to start giving individual loans to traders, ranging from $1,000 to $3,000 (❼ in Figure 5.12). Two traders decided to follow the recommendations and started receiving payments from retailers and wholesalers through the bank, giving the banks a record of their transactions. Others saw their success, and now 35 traders are using this new business approach.

Social and environmental innovations

Generating income growth in Mayurbhanj district was not just the result of financial changes such as gaining access to credit. Social and environmental initiatives initiated by UNIDO were essential to creating an environment in which loans and credit could work.

Insurance Personal accident insurance is now available for collectors, processors and traders, organized by a government agency with UNIDO's help (❽ in Figure 5.12). Collectors pay Rs 15 ($0.30) a year, and can receive up to $500 in the event of an accident. For processors the premium is $2 and the benefit up to $2,000. "Insurance camps" were organized in the villages to inform people about the insurance programme and get them to sign up.

Health training Many people in the area suffer from poor health, and large numbers of children never attend school, or drop out early. The self-help groups began raising awareness among local people about health, education and insurance services. They began providing help and advice on personal hygiene, arranging herbal gardens where families could grow medicinal plants; and prompting parents to send their children to school. They sometimes also produce lunch at schools and support with student tuition.

Forest conservation It is vital to conserve the forest for the future of the sal leaf value chain. Arable land in Mayurbhanj district is scarce, and there is a lot of pressure on the forest as a source of income: people fell trees for fuel and clear land to grow *sabai* grass. An earlier attempt by the Joint Forest Protection Committee to raise awareness about forest conservation and control illegal tree cutting had failed. But UNIDO's programme revived and reorganized this committee, and 40 women and 96 young people joined. A programme to raise awareness about forest protection targeted 95 villages, and trained collectors in sustainable

Table 5.4 Financial services in the sal leaf chain in India

	Loan from Ahari Federation to individual collectors (❷ in Figure 5.12)	Loans from bank to Betnoti Federation and processors (❺)	Loan from bank to collectors' and processors' self-help groups (❸ and ❻)
Product and financial flows			
Purpose	Short-term micro loan	Buying raw material and equipment	Loan
Amount	Loan given to individuals Total portfolio: $1,200	No cash given alone Raw materials, equipment and other items Total portfolio: $1,500	Total portfolio: $94,530
Period	7–30 days	6 months	At least 1 year
Interest rate	2% up to 7 days 3% 8–15 days 5% 16–30 days	1% for raw material 7% for equipment	11%
Transaction costs	1%	<1%	2%
Risk management			
Securitization	Peer pressure	Collateral: pressing machine Legal document stating terms and conditions	No
Liability	Both individual and self-help group	Individual	Self-help groups
Information flows			
Information required to apply	Application letter forwarded and recommended by self-help group	Application Federation membership	Bank account Savings Books and accounts Minutes of group meetings for at least 6 months

harvesting. To reduce illegal tree felling, a patrol squad was created, paid for by a fixed contribution from the 95 member villages.

Engaging banks with the sal leaf chain

In the past, there was a total mismatch between the sal leaf value chain and the financial system. Collectors received overly large loans and got into deep debt as they were unable to repay, while traders and processors had no access to bank loans and were unable to invest in their businesses. Now, after the intervention, the actors in the sal leaf chain have become clients for the banks. UNIDO's programme established two triangles of value chain finance:

- **Collectors – Ahari federation – bank** The collectors are organized and trained in business planning and financial management. They are informed of market trends and they have built assets through regular savings. In other words, the collectors have become knowledgeable businesswomen and as such they are attractive clients for the banks. The banks provide them with working capital, in accord with their needs.
- **Processors – Betnoti federation – bank** The processors also organized and got trained in business skills. They started saving regularly, on the basis of which they obtained bank loans for working capital. They learned to present their business case to the banks and get investment loans for new equipment. With these investment loans they can innovate and expand their businesses.

In addition, the banks also started to finance some traders. Two factors helped to set this up. One, traders applied more formal methods of financial accounting and could therefore present reliable data to the bank. Two, the banks got to know how the sal leaf chain operates and developed more confidence that it could recover loans from the sector.

Table 5.5 shows the risks of lending to the sal leaf chain. Various risks have been reduced thanks to UNIDO's intervention. Financial training, introduction of regular savings, and organization of the chain actors have been the key risk mitigation strategies.

Benefits

Collectors At the start of the programme, the collectors earned about $0.30 per day, or $10 a month. By selling directly to the Federation and not using the services of agents, they were able to increase their incomes to $14 a month.

Using the production loans from the banks, some collectors were able to buy a processing machine and increase their average incomes by $200 to $300 per month per person. This also gives them employment all year round, instead of just in the high season.

The collectors' self-help groups now meet regularly; the members are disciplined and have adopted a code of conduct. Three hundred groups have started group savings, 133 have started lending within the group, and 30 have negotiated with the bank to get loans that suit their needs.

Four hundred self-help group members now use their loans for business activities instead of consumption, and 240 group members have invested credit from internal lending in their business activities.

Processors The monthly income of processors has increased from $12 per month to some $35–$40 per month in the high season. This increase was the result of storing raw materials, improved processing technology and the introduction of paper as a raw material.

The Betnoti Federation supports processors by giving them access to raw materials in the low season. It gives them loans to install better technology and energy-saving devices. It also provides employment to many people. Some 120 processors are the direct beneficiaries of this initiative.

Traders Thirty traders received loans from banks and increased the scale of their business. They also benefited indirectly from the improved technology because it produces a better product. These benefits to the traders have a trickle-down effect through the whole sal production process in the area.

Banks Banks now experience better loan recovery: the repayment rate increased by 55%. The relationship between banks and lenders has improved thanks to hand-holding, which decreases their credit risk. More than 80% of the defaulting self-help groups improved their working relationship with the banks. In addition, the banks developed new client relations with the processors and the traders. The bank has learned that the sal leaf chain can be an attractive market for the bank.

The chain as a whole The chain as a whole has become more integrated through better information flows, better financial management and lower costs. New innovations in products and technology have added value to the chain. The scale of production has also increased, especially at the level of processors and traders. The value chain has been made more sustainable through conservation efforts.

Social and environmental impacts

Some 1,560 primary collectors and 120 processors now have personal accident insurance.

The introduction of paper into cup-making cut the use of sal leaves by half, supporting conservation efforts. The growth of the Joint Forest Protection Committee from 7 village members to 95, and its renewed activities has also had an impact: illegal tree felling has stopped in 47 villages because of the patrol squad.

The social impacts on the chain have also been significant. Most processors are women. Once they gained access to working capital, they could upgrade their equipment, buy cheaper raw materials, and earn more. Better technology has reduced their physical burden. Because they are earning more in less time, they now have more time for their families.

Traditionally, children were often involved in processing, meaning they missed school. As adults, they were less able to manage business activities and understand markets, making them overly dependent on others in the value chain. Once their mothers started earning more, the children had a better chance of getting an education and breaking out of this vicious cycle.

Table 5.5 Risk analysis for the sal leaf chain in India

	Source of risk	Risk level	Mitigation strategy
Production	Over-collection of leaves	●●●	Sensitization among collectors Forest protection regulations Alternative raw materials
	Energy inefficiency and losses	●●●	Improved pressing equipment
Price	Big changes in price of sal leaves	●●●	Increasing leaf storage capacity
Market	Competition from alternative products	●●○	Product development and diversification
Default	Misuse Non-repayment	●●●	Financial counselling Training on financial literacy Regular savings
Other	Illness	●●●	Insurance Health club for children and women Herbal garden

Risk levels: ●●● High ●●○ Medium ●○○ Low

Challenges and lessons

Length of intervention The UNIDO programme began in January 2006. In May 2008 the programme ended and UNIDO staff withdrew. Around 9,000 collectors, 200 processors and 30 traders benefited from the programme. As of February 2009, the Federations were still performing well. This programme demonstrated that it is possible for a donor-funded intervention in a value chain to achieve substantial and sustainable results, to hand over some minor continuing support activities to existing government agencies, and to close down in as little as 3.5 years.

Local agents The position of the local agents in the chain continues to raise a question mark. UNIDO did not involve them in the programme because they were adding costs and not value to the chain, and exploiting collectors and processors with excessive prices. An informal inquiry into the lot of five agents after the programme found that two had bought a pressing machine and started working as processors. The other three were still operating as agents in other villages (outside the 25 project villages). They have increased the price that they offer to primary collectors for sal leaves.

Credit The problems in the chain show that excessive value chain credit can be as damaging and ineffective as too little or no credit.

Multiple approach Successful value chain interventions must be multi-facetted, including inputs such as improving marketing and processing channels, building strong community institutions and technological changes. Value chain finance on its own can rarely achieve anything.

Importance of grassroots institutions The flow of value chain finance involves disaggregating large sums into the small amounts which individual producers, traders or intermediaries need. This requires strong and usually community-owned institutions, which must be built before any sustainable financial flows can follow.

Local adaptation Massive top-down government or donor credit delivery programmes must be radically changed to make them useful for the target group. The target group's needs and savings capacity must be the basis for the loans.

More information

Amit Kumar Singh, *Madhya Pradesh Rural Livelihoods Project, asinghks@gmail.com*

Sangeeta Agasty, *Foundation for MSME Clusters, sangeeta.agasty@gmail.com*

www.ruralenterprise.blogspot.com

Micro-factoring: Instant payment on delivery of tea in Kenya

Kenya is one of the world's top exporters of tea, and Kenyan tea has won international acclaim for its consistently high quality and pleasant aroma. Tea is also one of the country's top foreign exchange earners, and the Mombasa Tea Auction is the second largest in the world. About 60% of the country's tea is produced by smallholders, who earn an average of around $1,500 a year.

But Kenya's smallholder tea growers face a problem: they find it difficult to get paid on time. Some have to wait for months before they finally receive their money. The long delay makes it hard for them to feed their families in the meantime, or to invest in improving their farm enterprise.

Similar problems face producers of many other types of agricultural products: farmers have to invest up front, wait a whole season before harvesting the produce, then wait again for the buyer to pay. That limits their productivity and output, and means that they are forced to sell at low prices to intermediaries who can give them cash immediately.

Factoring as a solution

Biashara Factors Limited, the microfinance arm of Kenya Gatsby Trust, is developing solutions to this problem. It provides an innovative financing arrangement known as "invoice factoring" that offers short-term financial services to smallholders and other actors in the value chain.

Biashara has adapted this service for various commodities in Kenya: tea (mainly in the Kericho area in Rift Valley province), cotton (in the Siaya area in Nyanza province), fish (Lake Victoria), and horticulture, coffee and dairying (Central Kenya), as well as for information technology in arid and semi-arid parts of the country.

This case describes the factoring service that Biashara has developed for the tea value chain.

From bush to cup

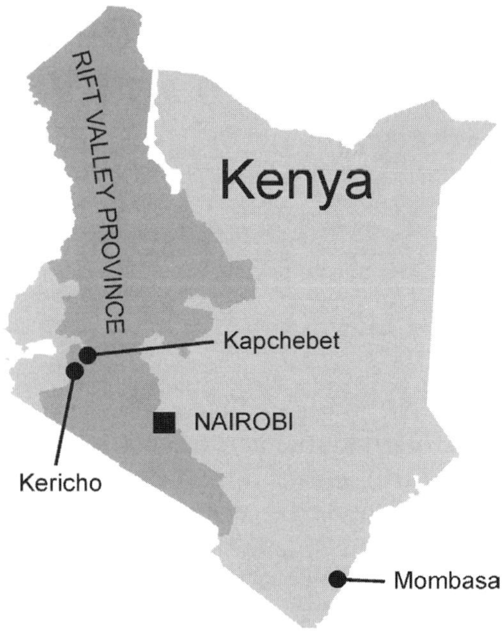

Figure 5.13 The Kabianga cooperative is near Kericho, Kenya

On the face of it, one would expect tea to provide growers with a regular income. During the picking seasons (March–May and August–December), the young leaves can be plucked from the tea bushes once every 2 weeks. The farmers take sacks of leaves to collection centres, from where they are taken to a processing plant for drying and processing (❶ in Figure 5.14). The processed tea is then taken to Mombasa for auction. Tea traders there buy the tea, pack it and sell it to the local or export markets.

Under this system, the Kenya Tea Development Agency (KTDA), a company that serves the country's smallholder tea growers, paid the farmers KSh 30 ($0.38) per kilogram of tea. But the farmers would have to wait up to 3 months before they finally got paid. So many farmers instead sold their output to private traders, who paid immediately – but much less: only KSh 10 ($0.13) per kilogram (❷ in Figure 5.14). The farmers were in a poor position to bargain because they needed cash urgently, did not know the prevailing market rates, and had no access to alternative sources of finance.

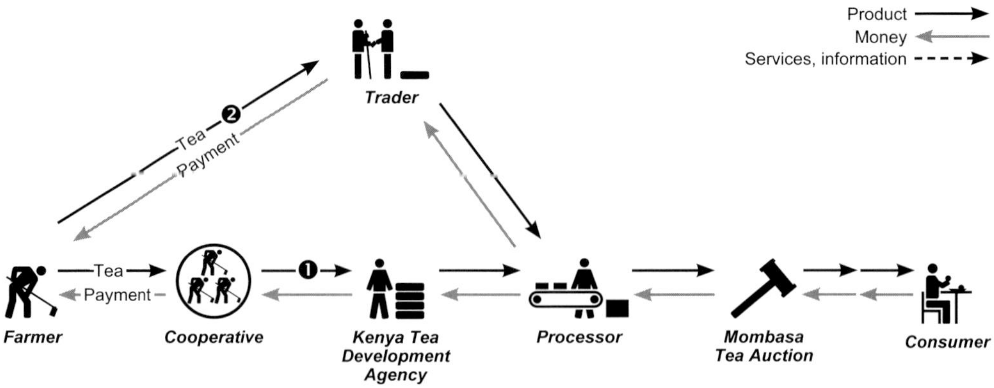

Figure 5.14 The tea value chain before Biashara's intervention

Box 5.3 What is invoice factoring?

Factoring is a form of business financing where you sell your invoices to a factoring company in exchange for immediate payment. It eliminates the 30–90 days that your customers take to pay your invoices, and provides you with the working capital you need to run your business. Factoring is a common practice for big business, but is revolutionary for micro-entrepreneurs.

This is how it works:

1 You deliver goods or services to your customer and issue an invoice.

2 You sell your invoice to a factoring company, which immediately advances you the first instalment, between 70% and 90% of the gross value of the invoice. You receive the first payment in as little as 48 hours.

3 The factoring company sends the invoice to the customer.

4 After 30–90 days, the customer pays the invoice and the factoring company pays you the remaining funds as a second instalment, minus the interest and a processing fee.

The beauty of invoice factoring is that you get predictable cash flow. It eliminates the uncertainty of when your customers will pay. The amount of money that is advanced depends on your sales volume, and your financing line increases as your sales grow. Factoring is easy to set up and is suitable for both established and new enterprises.

Building a new chain

The 3,000-plus members of the Kabianga cooperative, one of several co-ops in a tea-growing area near the town of Kericho in Rift Valley province, were convinced they could do better. Their cooperative collected tea from individual members, then sold it to KDTA. They decided to reopen a run-down tea processing plant at Kapchebet, which is jointly owned by individual farmers and the cooperative. The factory would sell the tea directly through the Mombasa auction, rather than going through KTDA's cumbersome procedures. They managed to obtain a loan of KSh 100 million ($1.25 million) from a national development bank to renovate the factory. This loan was secured by a charge on the factory's assets and the directors' guarantee.

The factory had to repay this loan on time. It also needed to guarantee supplies of tea from the farmers. That meant paying them quickly so they would not be forced to sell to the traders. The factory did not have enough working capital to do both, so the management approached Biashara for a factoring facility (Box 5.3). Biashara, a pioneer in providing this service to smallholders in East and Central Africa, is a self-financing organization that does not depend on donor funding for its operations. It gets its funds from its parent organization, Kenya Gatsby Trust, in the form of shares, and also borrows from other financial institutions in Kenya and foreign organizations such as Shared Interest, a socially oriented lender based in the UK.

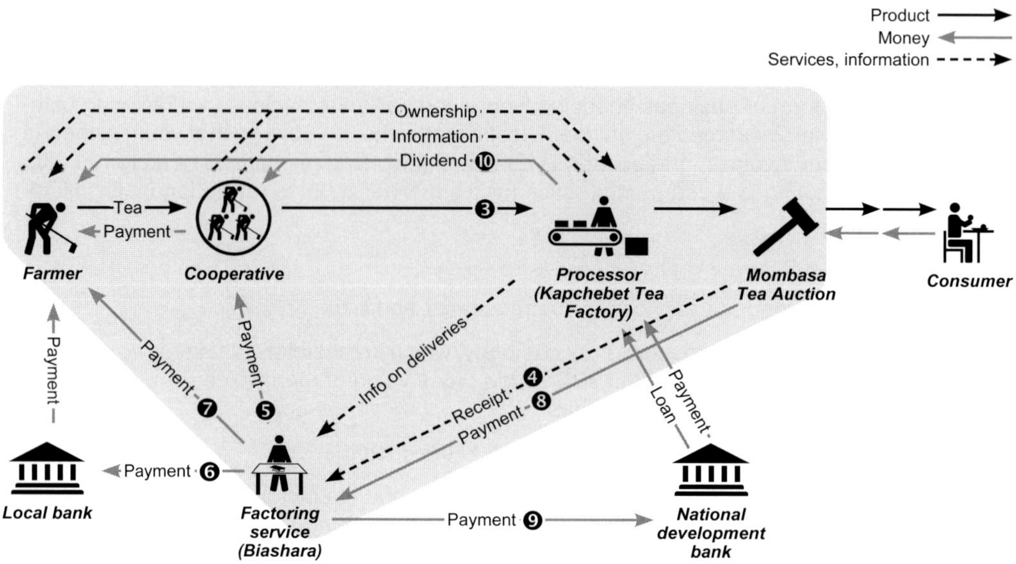

Figure 5.15 The tea value chain with Biashara's factoring services

Biashara conducted a comprehensive appraisal of the auction house to determine its financial and management rating, and studied the value chain to become familiar with the movement of the produce and payment flows. It analysed the various critical points in the chain, such as production, quality control (on farm and in the factory) and transport. It discussed the results of this study with the factory management and the cooperative, and helped them overcome issues that arose. After analysing the viability of the venture, it decided to start providing the factoring facility in phases to ensure that it worked well.

The resulting chain is summarized in Figure 5.15.

The farmers deliver their tea to the co-op's collection centres, as before. The co-op then transports the tea to

Figure 5.16 The factoring system uses Safaricom's M-PESA scheme, which lets farmers be paid via their mobile phones

the Kapchebet tea factory (❸ in Figure 5.15) which processes it and delivers it to the Mombasa auction.

Twice a week, the auction house sends a receipt to Biashara listing deliveries of tea it has received from the factory (❹ in Figure 5.15). The factory provides Biashara with a list of farmers and the amount of tea they have delivered. On the strength of this, Biashara pays out 70% of the money to farmers (or more, depending on the sales to the auction).

Making payments to over 3,000 farmers is a daunting task, which Biashara handles by paying into farmers' individual accounts with the co-op (❺), local banks (❻), or through the M-PESA or PostaPay money transfer services (❼ in Figure 5.15 and Box 5.4). These payments are made within 3 days of Biashara receiving the auction house's receipt. The tea delivered to the factory acts as security for these loans.

When it has sold the tea (about a month later), the auction house pays the full amount to Biashara – up to KSh 10 million ($125,000) a week (❽ in Figure 5.15). Biashara then deducts 10% of the total and pays this into the Kapchebet factory's

Box 5.4 Money transfer services

To remit small amounts to farmers, Biashara uses the M-PESA and PostaPay money transfer services.

M-PESA

M-PESA (M stands for mobile, while *pesa* is the Swahili word for money) is a service offered by Safaricom, a Kenyan mobile phone service provider, in partnership with Vodafone. It allows subscribers to transfer money using mobile phones, without needing a bank account (Figure 5.16).

When Biashara pays a farmer, an SMS is sent to his or her mobile to confirm the transaction. The farmer can then go to any one of Safaricom's 11,000 agents throughout Kenya and show the agent the phone number and an identity card, and say how much he or she wants to withdraw. The agent then pays the farmer this amount in cash. This service costs KSh 30 ($0.38) to send KSh 100–10,000 ($1.25–$125) to a registered M-PESA user, and up to KSh 75 ($0.94) to withdraw over KSh 10,000.

More information: www.safaricom.co.ke

PostaPay

PostaPay is a joint venture between the Kenya Post Office and Afripayments (the software company providing the service). Biashara sends a list of the farmers it wants to pay to Afripayments, along with a cheque for the money to be paid (including a service fee). The list includes the identity card numbers and nearest post office of each farmer. Afripayments distributes this list to the respective post offices. The farmers go to the post office with their identity cards and can withdraw their money. This takes a maximum of 3 days. The cost of transaction depends on the amount sent.

More information: www.postapay.co.ke

Table 5.6 Factor payments by Biashara to farmers for tea in Kenya

Product and financial flows	
Purpose	Provide prompt payments to tea farmers
Amount	Total limit of KSh 30 million ($375,000) Average of $25 per farmer per week
Period	30–90 days
Disbursement	Disbursed weekly, directly to farmers' bank accounts or through M-PESA and PostaPay services Initial payment = 70–90% of expected value of invoice Final payment = 10–30% of invoice, minus fee
Repayment	Repaid by the buyer (the Mombasa Tea Auction)
Interest rate	2.5% per month, deducted from the final payment received from the buyer
Transaction costs	One-off 1% payment on the estimated monthly volume
Risk management	
Securitization	Binding memorandum of understanding between the buyer, seller and the factoring house Tea processor may offer assets or produce as security Thorough assessment and investigation before offering financial services Buyer must pay factoring house rather than seller
Liability	Auction house and Kapchebet Tea Factory are jointly and severally liable
Information flows	
Information required to apply	For processing plant: • Application letter • Business registration certificates, value-added tax compliance, personal • identification number • 2–3 years of audited accounts • Current year's management accounts • Company and directors' profiles, copies of directors' identity cards or passports • Company organizational chart • Sale contracts between factory and auction house, farmers and cooperative For farmers: • Account number, co-op number, ID number and location
Information required during season	Monthly reports by field officer of Kenya Gatsby Trust to financiers on impact of funds: reports from processing plant on deliveries, auction house receipts, number of farmers reached
Time lag between application and payment	5 days from appraisal to background check Disbursement immediately after approval

bank account to cover its processing services (❾). The bank deducts the loan repayment instalment from the factory's account. Biashara then pays the farmers the balance of what is due to them (again following arrows ❺, ❻ or ❼), minus 2.5% interest per month. This interest charge covers Biashara's costs. Biashara currently finances an outstanding balance of about $130,000 between the auction house and the farmers.

To ensure that the farmers fully understood the factoring process, Biashara offers quarterly training courses for farmers on basic bookkeeping and the importance of sustainable business relationships, and helped the cooperative set up systems to support the process. It covers the cost of these services from its fees.

Table 5.6 summarizes Biashara's factor payments made to farmers.

Risk analysis

Tea is a risky business with a fragmented market structure and strongly fluctuating market prices. Nevertheless, Biashara has managed to develop a financing model which effectively mitigates the major risks. The key elements are (Table 5.7):

- **Triangular cooperation** Before starting to finance the chain, Biashara conducted extensive research to understand the workings of the chain and to check the conditions of the companies in it. This phase of research and due diligence results in a contract agreement between the suppliers, the buyer and Biashara for the delivery of the factoring service. This agreement details how information flows repayment will work, and outlines each party's roles. The agreement forms part of the collateral for the financing.
- **Lead firm model** Though small farmers receive the factoring service, the repayment is done by a large lead firm (the auction house). Hence, Biashara's risks are not with the small farmers, but with the lead firm. Therefore, in the research phase, the due diligence focuses especially on the financial condition of the lead firm.
- **Partial pre-finance** The international market price of tea fluctuates, so Biashara cannot know in advance how much to pay the farmers. It mitigates this risk by estimating the price beforehand and paying farmers a percentage of the price as a first instalment. When the actual price is known, Biashara pays the amount remaining, minus interest and fees.
- **Alignment of interests** The farmers have an interest in selling through the Kapchebet processing plant because they hold shares in it: if it earns a profit, they earn a dividend (❿ in Figure 5.15). They have an interest in the factoring system because it pays them quickly and pays higher prices than they would get from the alternative – selling to private traders. The cooperative maintains good relations with the farmers by providing them with various services, including extension advice and fertilizers on credit.

Table 5.7 Risk analysis for tea factoring in Kenya

	Source of risk	Risk level	Mitigation strategy
Production	Seasonality	●●○	Increase buffer stock during high seasons
	Disease	●○○	Train farmers on good production methods
Price	Low demand	●○○	Market forces and constant marketing
	Market price fluctuations	●●○	Encourage farmers to diversify
Default	Side selling of produce	●○○	Close relationship with farmers Prompt payment Farmers own shares in factory
Currency	Currency risks on international loans	●●●	Minimal borrowing from abroad

Risk levels: ●●● High ●●○ Medium ●○○ Low

Benefits

Farmers

The 3,000 farmer-members are the main beneficiaries of the factoring system. They get prompt payment after delivery and higher prices than they could otherwise expect: KSh 30 ($0.38) per kilogram of tea, compared to KSh 10 ($0.13) previously offered by the traders. This has reduced the number of farmers seeking the services of independent traders. A few farmers divert their produce to other buyers, but their numbers are not significant.

A culture of saving and borrowing is growing among the farmers, and farmers are able to leverage their savings to access other financial products.

The farmers are able to make informed decisions about their production because they can now obtain information directly from the cooperative and the factory. The factory publishes accounts that show how much it has earned and how profitable it is. The cooperative is able to negotiate better terms for its members because they are corporate shareholders in the factory.

Cooperative and Kapchebet factory

With fewer members selling to other traders, the co-op and factory have increased the volume of tea they handle. They have improved their credibility and reputation, and are able to attract new members.

The cooperative now is in a position to approach other financial service providers to develop specific products for its members.

The tea factory's sales have grown from 5,000 to 8,000 tons a day. It has a guaranteed supply of raw materials and is able to plan its processing schedules with minimal risk of its suppliers diverting their produce elsewhere. It has been able to repay most of its initial loan, and plans to apply for a second loan to expand its operations.

Value chain

Bottlenecks have been removed from the value chain as a whole. A greater volume of produce now moves more smoothly than before, and payments are made on time.

The strong linkages between the different actors cultivates a culture of responsibility in the chain.

The chain provides a clear communication channel, keeping everyone informed about pricing and the quality required.

Challenges

- **Sensitization** Factoring is a new financial service in Kenya, and many people shy away from it. This is not because of its complexity but due to a lack of knowledge. Many people view factoring as a kind of loan. Biashara tries to educate the public by organizing forums as part of its marketing work.
- **Complexity of the groups** A lot of effort is needed to mobilize the producers into business groups before the factoring service can be introduced. This is costly and time-consuming. Most financial institutions do not have the patience to do this groundwork; they rely on NGOs to establish a relationship with the group.
- **Lack of regulation policy** Factoring can be abused. It is not regulated by banking laws, and so unscrupulous operators could use it to defraud clients. Biashara is pushing for more factoring houses to be set up to have enough weight to lobby for regulations on aspects such as capital base, dispute-settlement procedures and licensing.
- **Lack of investors** Because factoring is new to Kenya, potential investors lack information, so are unwilling to put money into this form of financial service.

Lessons

- Factoring can be used to alleviate smallholder farmers' from cash-flow problems. Such farmers lack hard collateral to secure finance from banks and other financial institutions. The factoring invoice provides security that enables the farmers to obtain funds.
- Factoring complements other innovative services such as M-PESA and Posta-Pay money transfers. These enable financial services to reach large numbers of widely scattered farmers who lack bank accounts.
- Factoring is flexible enough to be easily replicated to other commodities.

- Factoring builds capacity of farmers and other actors and strengthens the value chain. Policy makers should develop legislation to regulate and promote the industry.

More information

Billy O Mathu*, Biashara Factors/Kenya Gatsby Trust, email bmathu@kenyagatsby.org, bllymathu@yahoo.co.uk, websites www.kenyagatsby.org, www.biasharafactors.com*

Financing rice farming in Rwanda

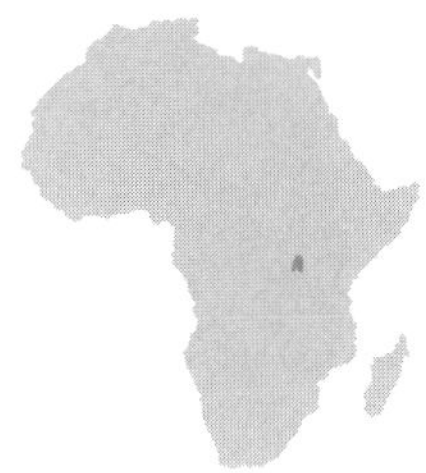

RICE GROWING HAS A short but eventful history in Mukunguli, 75 km south of the Rwandan capital, Kigali. In the 1970s, Chinese entrepreneurs leased marshland in the area and turned it into irrigated rice fields. They encouraged local farmers to plant rice, and for over 20 years held a near-monopoly over all the rice they produced. They bought most of the paddy rice. The farmers made little money, but it was still better than nothing.

Struggling to raise their output, the farmers turned to local moneylenders for loans so they could buy inputs. They promised to repay the loans in the form of dried paddy. The interest rates were high: sometimes the repayments amounted to three times the original loan. The moneylenders took around 50% of the farmers' rice. Another 20% went to the Chinese in return for use of the land. The farmers used 10% for home consumption and sold the remaining 20% to local shopkeepers.

After the genocide of 1994 the Chinese fled the country, as did the farmers and many others. The farmers later returned and started to cultivate rice again, this time without the Chinese. A German NGO helped rehabilitate the irrigation system, but finding a market was difficult, as was getting fertilizer. Low yields forced some farmers to start growing sweet potatoes and beans instead of rice.

Reviving the rice chain

But most of the farmers did not give up on rice: they knew that they could benefit by coming together to bulk their rice, negotiate better prices and seek new markets. Plus, being organized would grant them better access to inputs, extension services and credit. It took almost 10 years to set up a farmer organization, but in 2003, 280 rice growers established the Rice Producer Cooperative (COPRORIZ). The result was the value chain shown in Figure 5.18.

Figure 5.17 Location of Mukunguli, Rwanda

Farmers The farmers are smallholders who earn between $200 and $500 a year. They grow and harvest the rice, dry it and take it to the co-op. Currently there are over 5,000 farmers in the area who grow rice, owning a average of 0.4 ha of rice land each.

COPRORIZ This co-op provides its members with a number of services. Before and during the season it supplies inputs such as fertilizers, seeds and pesticides. At harvest, it collects and bulks their rice, takes it for milling, and sells it to traders or wholesalers (❶ and ❷ in Figure 5.18). It also provides extension and training. The co-op's general assembly, composed of all members, meets twice a year and elects a board of directors. A manager runs the co-op from day to day. Almost all the area's rice growers are members of the COPRORIZ; the few non-members are still allowed to sell through the co-op.

Crop traders and wholesalers Two types of traders buy rice: crop traders buy unmilled paddy, while wholesalers deal only with milled rice. The co-op prefers to sell to the former as it can avoid the costs and delay involved in milling and transporting the rice. But if the traders offer a price that gives the co-op less than a 25% profit, the co-op will arrange for the rice to be milled and delivered to a wholesaler.

Millers The millers mill the paddy to make white rice ready for cooking. A number of small mills exist, but they produce a low-quality product and a lot of wastage. The co-op prefers to have its rice milled by larger mills in Kigali, which have less milling loss and make it easy to deliver direct to wholesalers in the city. However, this involves transporting the paddy from Mukunguli to Kigali, waiting for the rice to be milled (the mill serves many producers from different parts of the country), and paying a service fee of $60 per ton of white rice, payable immediately.

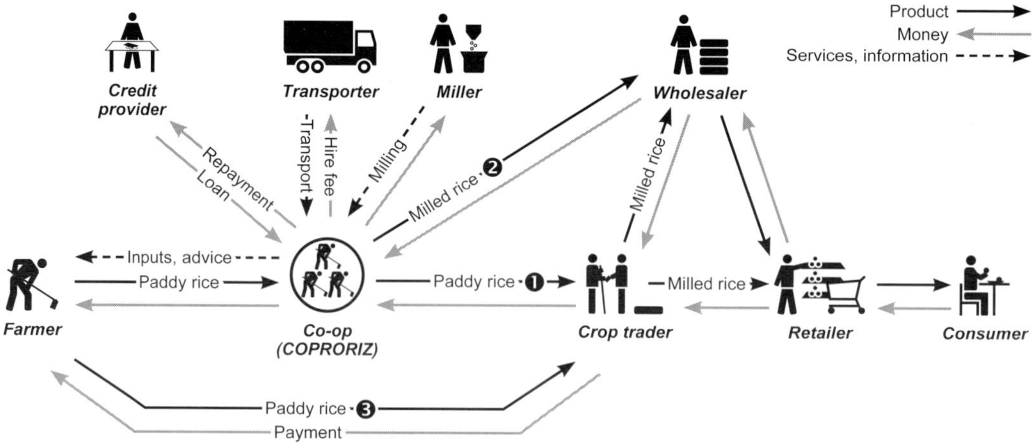

Figure 5.18 The initial COPRORIZ rice value chain

Retailers Shopkeepers and supermarkets buy the milled white rice from the wholesalers and sell it in packages of 1–5 kg. The locally produced rice competes with rice from Thailand and Tanzania.

Consumers The consumers include residents of the Mukunguli region and Kigali.

Problems in the chain

This chain was not without problems:

Lack of inputs The co-op bought inputs such as seed and fertilizer in bulk, but could not distribute them to farmers on credit as it lacked the working capital to do so. Strapped for cash themselves, the farmers could not afford to pay for them even at the reduced prices the co-op charged. So many farmers would plant without fertilizers, and would harvest less than 3 tons per hectare instead of the 5 tons/ha they could expect with fertilizer. Some would delay planting until they had enough money to buy inputs. Staggered planting allowed pests and diseases to spread from one field to another, and different harvest times complicated marketing and transport.

Slow payment Once the rice was safely harvested, dried and delivered to the co-op, the farmers still had to wait for up to 2 months to be paid because it took that long for the buyers (in Kigali) to pay the co-op. In need of cash, many individual farmers started to sell their paddy directly to local traders who paid only one-third the price ($0.33 per kg, compared to $0.90 from the co-op), but handed over the money immediately (❸ in Figure 5.18). By 2007, only 40% of the total rice produced was marketed through the co-op.

Loss-making co-op Although the co-op served many needs of its members, it was making a loss. For income, it depended on a membership contribution of $2, and on the fees it charged for a range of services: the sale of rice, provision of fertilizers and seeds, the provision of sacks and organizing transport of paddy to mills, arranging extension services, and the like. The level of these fees was fixed by the general assembly; they typically covered the costs incurred plus a management fee of 30%. But this income was not enough to cover the full costs of the co-op.

CAF Isonga

A local microfinance institution, Caisse des Affaires Financières (CAF) Isonga, had experience in financing rice production in the Gitwe region in northern Rwanda. CAF Isonga was looking for new opportunities, and it established a branch in Mukunguli. It had heard of the co-op's problems and offered to help overcome them.

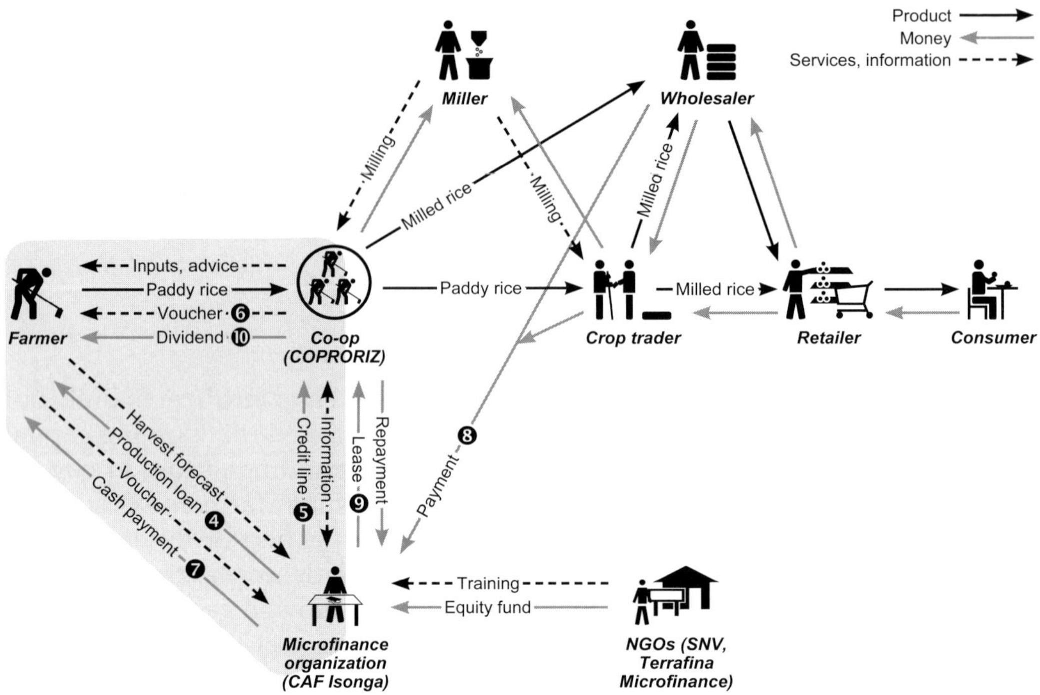

Figure 5.19 The CAF Isonga/COPRORIZ rice value chain

Starting life as a savings-and-credit cooperative, CAF Isonga was transformed into a limited liability company in 2006 so as to meet central bank regulations for microfinance providers. With an equity of $375,000, the company has assets of $1,327,000 (in December 2008), and a clientele of over 21,000; 84% of these are smallholder farmers, its main target group. CAF Isonga is the only formal financial institution to serve the rice value chain in Mukunguli. It has become the main source of finance for COPRORIZ and its members, and has established a branch near the rice farms to cater for the financial needs of the rice producers.

Smoothing the financial chain

CAF Isonga has developed a variety of financial services for the farmers and the co-op. To do this, it received technical support from SNV (the Netherlands Development Organisation) to develop its services, and from Terrafina Microfinance, which provided training and equity capital. These financial services involve close collaboration with the co-op and the farmers.

Production loans CAF Isonga decided to provide credit to farmers so they could buy fertilizer and hiring labour. Only farmers that are members of COPRORIZ can apply for these loans; the co-op screens each applicant for integrity and capacity to repay – as evidenced by the size of the applicant's rice plots. It guarantees the loan by co-signing the contract between CAF Isonga and the farmer, agreeing to repay the loan if the farmer defaults. If the application is approved, CAF Isonga transfers the money to the borrower's bank account (CAF maintains accounts for all the co-op members, which farmers can use for savings as well as to manage their loans) (❹ in Figure 5.19). The farmer repays the loan by delivering paddy to the co-op (see the voucher system below). If the farmer defaults on the loan (for example, by not delivering to the co-op), the co-op has to repay the debt. The average loan is $100; the loan conditions are summarized in the second column of Table 5.8.

Paddy commercialization loan To enable the co-op to pay farmers more quickly, CAF Isonga developed a "paddy commercialization loan". This is a credit line that allows the co-op to pay farmers on the same day that they deliver rice to the co-op warehouse (❺ in Figure 5.19). The co-op bulks and stores the rice until it is a good time to sell. Once it has found a buyer prepared to pay a good price, the co-op takes the rice to a miller and delivers it to the buyer. It then repays the loan, plus interest, to CAF Isonga. The ceiling of the credit line depends on the estimated rice output for that season. In 2008, it was set at $464,000. The third column of Table 5.8 gives the key features of this credit line.

Voucher system A voucher system is used to speed payments to the farmers. When a farmer delivers rice to the co-op warehouse, he or she is given a voucher showing the amount, co-signed by the warehouse manager and the CAF Isonga credit officer (who doubles as warehouse keeper) (❻ in Figure 5.19). The co-op can give out vouchers up to the maximum amount of its credit line. The farmer presents the voucher to the CAF Isonga office just down the road from the warehouse, and CAF Isonga pays the farmer the full value of the paddy delivered, after deducting the production loan and interest (❼). The value depends on the price the co-op has negotiated with the trader (it sells to the trader who offers

Figure 5.20 The combined efforts of the co-op and microfinance institution mean benefits for farmers

Table 5.8 Key features of CAF Isonga's loans to farmers and COPRORIZ for rice in Rwanda

	Production loan for farmers (❹ in Figure 5.19)	Paddy commercialization credit line for co-op (❺ in Figure 5.19)
Product and financial flows		
Purpose	To enable rice farmers to pay for inputs such as labour, pesticides and fertilizers	Pre-finance of purchase and marketing of paddy rice
Amount	Total $286,000 Average loan to farmer: $100	$464,000
Period	7 months	3 months
Disbursement	Disbursed to borrower's bank account just before planting	Ceiling for credit line set through agreement Payments directly to farmers upon submission of signed vouchers showing amount of rice delivered to co-op warehouse
Repayment	Deducted directly from proceeds of sales to co-op just after harvest (March and September)	Repaid by co-op after rice sale through direct deduction from sales proceeds
Interest rate	1.5% per month	Negotiable, depending on ceiling and loan term Varies between 2 and 2.75% per month
Transaction costs	$1 per application plus 1% commission	1% commission fee
Risk management		
Securitization	100% guarantee by co-op All transactions carried out through bank account at CAF Isonga, with facility to deduct amount due automatically	Warehouse inventory, joint warehouse management, and joint marketing of rice
Liability	Individual farmers and co-op are equally liable	Co-op is liable
Information flows		
Information required to apply	Updated yield forecast Approval from co-op, specifying amount needed and production capacity Credit history	Production forecasts Prices Potential buyers Needed amount (ceiling) based on production forecasts
Information required during season	Price of agricultural inputs Price of rice	Monthly reports on disbursements
Time lag between application and payment	2 days	1 week from application to agreement

to pay most). The trader pays the co-op through CAF Isonga (❽), enabling the co-op to repay its credit line.

This system works well because all payments are made through CAF Isonga, and both the co-op and CAF Isonga are custodians of the warehouse and jointly control the flows of paddy into and out of it. The repayment rate is 100%, and the portfolio at risk is close to zero (Box 5.5).

Lease for transport CAF Isonga's credit officers helped COPRORIZ carry out an internal assessment to find out why the co-op was continuing to lose money. They discovered this was mainly due to transport expenses. The co-op used to hire lorries to carry paddy to millers in Kigali; the costs were so high that it was

Box 5.5 Callixte Niyonsaba talks about vouchers

"I am a rice farmer in Mukunguli, Rwanda. I own 2 hectares of land, including 0.7 hectares of rice. In the beginning, the rice plots belonged to the Chinese who started experimenting with this crop in Rwanda in the early 1970s. The Chinese allowed us to use the land, and in return asked us to give part of the production to them.

"We worked for nothing: we still owed money to moneylenders, who charged very high interest rates. Around 50% of our production went directly to moneylenders to repay loans. We also ate some of the rice ourselves. We sold the rest to local traders at very low prices – but at least the traders paid on the spot.

"In 1994, people fled genocide and war, and the area was left uncultivated for quite some time. People started growing crops there again in 2003. That year I got hold of a plot which I was allowed to cultivate, and together with almost 300 other farmers we decided to set up a cooperative, called COPRORIZ.

"Being members of the co-op has changed a lot for us rice farmers. Instead of us having to go to moneylenders to buy inputs, now the co-op was supplying fertilizer and seeds. And COPRORIZ has an arrangement with a microfinance institution, CAF Isonga. It gives loans to co-op members. Now it is easy: when the planting season is on, you go to the co-op, which gives you a guarantee for CAF Isonga, and it gives you a loan straight away. Now we no longer hesitate to get a loan, so we have been able to increase our production. We can also use the loan to cover household expenses. In return for the loan we have to sell all our rice to the cooperative.

"There is another arrangement that has helped us: the voucher system. Now we are paid upon delivery!

"The new situation has changed a lot. We work for ourselves; we have access to inputs, which helped increase our production. We are owners of the cooperative. If any profit is made, it is for the farmer. For example, I have been able to buy bicycles so other people can look after my farm when I am not around, and a motorbike and a mobile phone. I have even built a house. Before I started rice farming, I had none of these, and have acquired them gradually. I used money from rice to expand to other business as well: I have now three cows producing 30 litres of milk every day, and have started cultivating cassava and sorghum.

"The co-op has been crucial for this success. Without it we could not have had access to microfinance. COPRORIZ gave CAF Isonga confidence to lend to us."

impossible to make a profit. The cooperative was too new to be able to buy its own lorry, and it had nothing to use as hard collateral to get a loan (the warehouse and other buildings it used actually belonged to the Chinese). So CAF Isonga leased a 5-ton truck to the co-op (❾ in Figure 5.19), and it started to generate a net income.

Why a lease rather than an investment loan? Because CAF Isonga still owns the truck until the co-op has paid off the lease. That reduces CAF Isonga's risk. The repayment was supposed to be complete by 2008, but the contract was extended until June 2009 because heavy rain destroyed part of the crop and affected the co-op's cash flow. Previously, in 2007, rehabilitation work on the irrigation system also reduced production levels.

Profits and dividends. Thanks to lower transport costs, better governance and transparency, and fewer production leakages, the co-op started making a profit in 2008. The co-op retains around 10% of its profits as savings and then distributes the remainder to its members as a dividend (❿).

Table 5.9 Risk analysis for the rice chain in Rwanda

	Source of risk	Risk level	Mitigation strategy
Production	Weather	●○○	Establish a branch of CAF Isonga near the co-op and monitor forecast and actual production closely Monitor the difference between forecast and actual harvest to guide decisions about loan products
	Warehousing	●○○	Joint management of warehouse between CAF Isonga and co-op Daily management of inventory
Price	Price fluctuations	●●○	Help co-op market rice
Market	Transport	●●●	Lease lorry to co-op
	Demand fluctuations	●●○	Help co-op market rice
	Competition with imported rice	●●●	Lobby co-op ensure high-quality milling
Default	Over-indebtedness of borrowers	●○○	Base loan decisions on credit history and production forecasts
	Deliberate default	●●○	Co-op screens borrowers and assesses applications Co-op and farmer jointly liable for repayment
Other (governance)	Cooperative management	●●●	Joint decision-making between CAF Isonga and co-op Signed contracts
	Members' participation	●●○	Regular meetings with co-op members

Risk levels: ●●● High ●●○ Medium ●○○ Low

Mitigating risks

The value chain includes various financial risks (Table 5.9):

Production risk Rice depends on having the right amount of water, so if the weather is dry, or if a storm damages the irrigation system, production can be impaired. Rice may also be damaged in storage.

Market and price risk Prices and demand for rice fluctuate a lot. Wholesalers may prefer to buy imported rice if this is cheaper. The voucher system depends on the traders' willingness to buy the rice and the price they offer. These are the result of long negotiations, but no formal contract is signed, and the traders may drop out of the deal at the last moment. If they do not buy, the co-op will be in trouble, since it will not be able to repay its credit line.

Default risk Farmers face many demands for cash: production, regular family needs, and emergencies. These competing demands may lead to default through side-selling. The farmers' ability to repay depends on many factors: the market price, transport and milling costs, the availability of inputs, and so on. If the co-op does not manage to sell all its rice at an acceptable price, the farmers may default on their loans.

Governance risk The whole system depends heavily on the cooperative. This is well-managed, but in the future, poor accountability, inadequate transparency, and low levels of participation of farmers in decision-making may threaten this.

Benefits

Farmers The production loans allow farmers to buy fertilizer and hire labourers to work their fields, so increasing yields. Productivity rose by around 30% between 2007, when production loans were introduced, and the end of 2008.

The farmers get paid immediately on delivering the rice to the co-op warehouse. The farm-gate price of paddy has risen by 67%, from $0.33 to $0.55 per kilogram over the same period.

Higher and more reliable incomes enable the farmers to invest in new equipment: 95% now own a bicycle (compared to only 20% in 2005); 3% have bought a motorbike, and three have their own small-scale rice mills. Some have bought more land. The farmers are more organized financially: all now have a bank account, compared to only 3% in 2003.

COPRORIZ The co-op has a credit line that enables it to buy grain from farmers and hold it until market prices are attractive. Members have stopped selling to other traders, and the co-op retains 10% of the profits as savings. It has cut costs through leasing a lorry, and has professionalized its management and operations. Until 2007 it recorded losses, but in 2008 managed to post a profit of $29,500.

CAF Isonga The microfinance institution has expanded its operations and developed a reliable, profitable set of clients in a new area. The repayment rate is 100%. In only 4 years, its agricultural loan portfolio has increased 18-fold. In 2007, it replicated the product in other parts of the country. Mukunguli currently accounts for 32% of its agricultural loans.

Value chain The system has worked so well that the government is pushing for it to be replicated in other areas and on other commodities. The Ministry of Agriculture has asked CAF Isonga to support other microfinance institutes to develop similar services in various parts of the country.

Challenges

Costs Production costs are still high; it is expensive to transport paddy to Kigali, and the few new, small-scale rice mills are inefficient. The co-op is planning to acquire its own mill nearby, and expand the area it covers to include other rice-producing areas. CAF Isonga has agreed to help raise the resources needed for this project.

Governance The majority of co-op members are illiterate and cannot adequately monitor the organization's operations and ensure its leadership is accountable.

The co-op is a vital player in the system – which makes it very fragile. In addition, permanent deals have yet to be agreed with wholesalers and retailers to ensure that the chain is sustainable.

The co-op board often rubber-stamps the decisions of the manager, jeopardizing the organization's governance. There is need to ensure that the co-op is able to manage its many roles and at the same time respond to its members' needs.

Expanding financial services CAF Isonga considers the rice value chain as its flagship. It faces challenges in increasing its portfolio to match the demand for its services.

Lessons

- To be effective, value chain finance needs to be coupled with chain empowerment: the chain actors need to be able to take their own decisions, based on good information and knowledge, and be better organized so they can defend their interests.
- Soft collateral can work as alternative guarantee mechanism for chain actors to access finance.
- Access to finance is important in chain development. Integrated efforts are needed to develop agriculture-based value chains, especially by the private sector. Value chain finance works better when it is embedded in a holistic process of market, institutional and organizational development for chain

actors. This requires a strong partnership between chain supporters (such as CAF Isonga) and chain actors.

- Value chain development can work with little or no outside financial investment. When appropriate financial products are made available, the chain actors have the capacity to invest themselves. Even smallholder farmers are bankable. The finance provider may have to walk an extra mile to support farmers integrate into the chain. This requires a thorough analysis, good preparation and alternative collateral systems to make farmers' access to finance a success.

Further information

Kabundi Joseph, *CAF Isonga SA., jkabundi@yahoo.fr; caf-isonga@yahoo.fr*

6 Unleashing investments in the chain

IN OUR FINAL SET of cases, we see how financial agents enable chain actors to make medium-term investments through a range of financial services beyond the traditional, short-term loans.

- **Financing the honey chain in Kitui, eastern Kenya**, for example, describes how various branches of the K-Rep Group, a microfinance organization, stimulated honey production and marketing by micro-leasing equipment and by providing loans and factoring services.
- **Financing the soybean value chain in Ethiopia** describes how FCE (an NGO) and Harbu, a microfinance institution it is affiliated with, have promoted the processing of soybean into milk. Services have included seed capital, material support, technical capacity building, loans for investment and working capital, and leases of processing equipment.
- **Opening opportunities to small coffee producers in Nicaragua** discusses the way that the government and FDL, a microfinance institution, offer a series of grants, investment loans and working capital loans to individual farmers and a farmers' cooperative.
- **Developing the organic quinoa chain in Bolivia** illustrates the idea of a temporary joint venture, in which the financial agent and a business each invest money for a limited period in a joint business activity. The business provides the technical expertise, while the financial agent provides not only much-needed additional funds, but also its financial expertise.
- **Financing artisanal fishing in India**, the final case, is a fascinating account of how SIFFS, a federation of fishermen's and -women's societies, has reorganized the marketing of fish. This reorganization benefits the fishermen and -women, rather than traders, who used to take a disproportionate share of

the profits. Through its control of the marketing, SIFFS can access funds from banks, finance houses, insurance companies and donor organizations to offer its members investment loans, loans to help members repay existing debts that charge exorbitant interest rates, insurance, working capital, and a whole range of non-financial services.

Financing the honey chain in Kitui, eastern Kenya

BEEKEEPING IN KITUI IS almost as old as human settlement in this part of Kenya. Honey from Kitui used to be well-known: for many years the Tana and Athi Rivers Development Authority (TARDA), a regional government agency, processed it and guaranteed a market for beekeepers. But in 2003 this service was run down because of political interference, and the processing unit was moved to another area. Without a viable market, local beekeepers could no longer make a profit, and the quality of their honey deteriorated. Lower quality meant lower prices: while TARDA used to pay KSh 100 ($1.25) per kilogram, traders would pay only KSh 30 ($0.38). Rather than selling at such prices, many producers consumed their honey at home, sold it to local brewers, or gave up producing altogether. For many producers, honey was their second most important source of income (after goats), but even when prices were good, many survived on less than $0.50 a day. Their situation had become desperate.

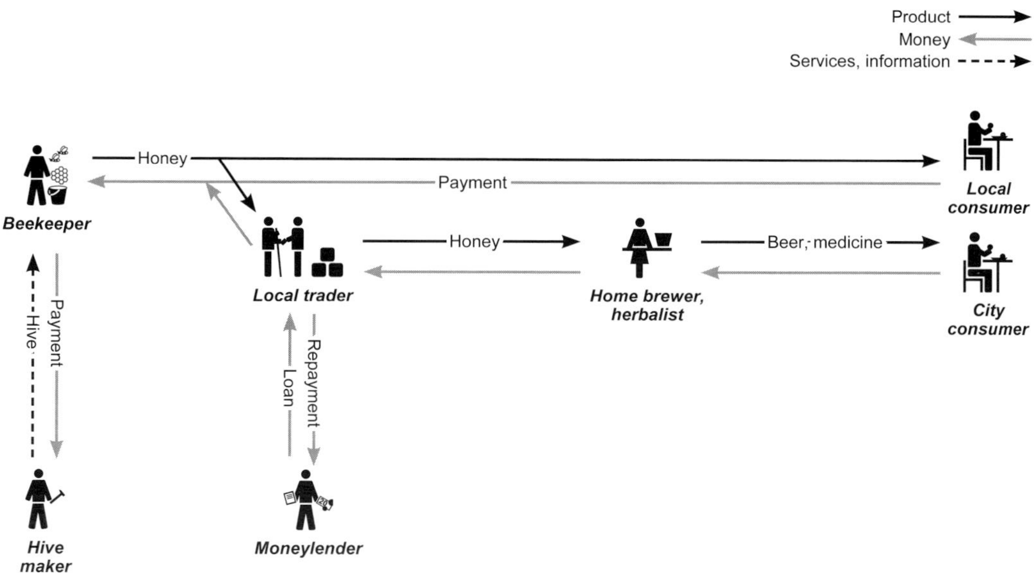

Figure 6.1 The value chain before the intervention

Figure 6.2 Traditional hives are made of hollowed out logs. Producers bottled and sold raw, unprocessed honey

Figure 6.3 Improved equipment enables producers to use improved hives and bulk their honey

The traditional honey chain

Relatively few actors are involved in the traditional honey chain (Figure 6.1). Support services like finance, technical assistance and business development are poorly organized or absent altogether.

The chain actors include:

- **Artisan hive makers** These hollow out logs to make hives and give them to beekeepers in exchange for livestock, cereals or honey. Only a few people make such traditional hives as it is not a profitable business.
- **Beekeepers** Almost every man in Kitui owns a hive, but only a few have bee colonies. Because of the low prices of honey, many have given up their hives. They have inadequate production skills and are not aware of market requirements.
- **Local traders** They buy small quantities of low-quality honey from beekeepers.
- **Home brewers and herbalists** Brewers buy bulk crude honey from the traders and brew it into a kind of beer. Herbalists use the honey to make herbal medicine.

- **Consumers** There are two types of consumers: local consumers in Kitui, who buy honey from neighbours, and consumers who buy beer and herbal medicine in Nairobi.

There used to be no formal financial services. Moneylenders lent money to local traders and charged high interest rates. Various microfinance institutions offered credit, but these were not aligned with the needs of the beekeepers and honey traders.

Figure 6.4 *Kitui honey is now better quality, so supermarkets are willing to sell it*

Reviving the honey business

Two studies in 2005 and 2006 funded by Danida (the Danish International Development Agency) found that Kitui district had over 400,000 traditional log hives, along with 3,000 modern hives with movable frames (which make it easier to harvest the honey). But the district produced only 3,000 tons a year, way below its immediate potential of 10,000 tons. Producers used poor beekeeping techniques, and charcoal-makers indiscriminately felled trees, whose blossoms were a food source for the bees. Production was far below demand, and packers had resorted to importing honey from southern Sudan and Tanzania. In the medium term, the researchers found, the honey subsector might produce up to 50,000 tons a year. The market value would be KSh 3 billion ($37.5 million) if the chain were efficiently organized.

The studies called for three interventions to revive the chain: form strong producers' organizations, strengthen market linkages, and provide appropriate financial services. Danida's Agricultural Business Development programme contracted K-Rep Development Agency (KDA) to pursue these initiatives. KDA is a research and development agency that focuses on developing, testing and refining appropriate financial products for marginalized groups in Kenya. It is part of the K-Rep Group, a microfinance organization focusing on poverty alleviation in Kenya (Box 6.1). KDA took on a coordinating role in reviving the honey chain, and called on other units in the K-Rep Group for support.

Designing a new chain

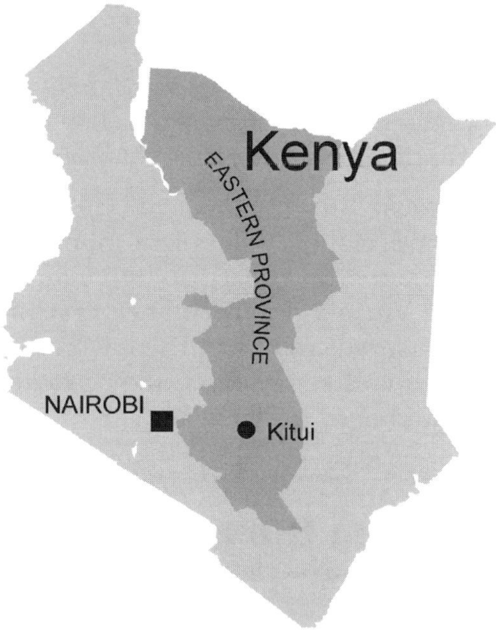

Figure 6.5 Location of Kitui, Kenya

KDA reviewed the demand for services by each of the chain actors, then designed a series of financial and technical services to build a new chain. With Danida's Agricultural Business Development programme, farmers were mobilized into groups and were trained on the various aspects of group dynamics and the basic principles of beekeeping as a business. Farmers with previous rudimentary skills in hive management were identified for further training to qualify as providers of this service. The existing local traders were selected and recruited into the programme. It was realized they lacked market information and record-keeping skills, among other things.

Baraka Agricultural College, a training institute based in Molo, in Kenya's Rift Valley province, trained the hive makers, producers and traders on production techniques. The traders in turn train groups of producers on how to maintain their hives and improve the quantity and quality of honey.

KDA trained the producers and traders on financial management and business development, and built the capacities of staff and board members of the financial services associations (see below).

The new chain began operating in 2007 (Figure 6.7).

Hive makers The KDA project trained the hive makers how to fit the traditional log hives with a "queen excluder" – a mesh that stops the queen from laying eggs in the honeycombs, so preserving the honey. This is a much cheaper option than modern hives, which were being widely promoted, even though they are not necessary to produce good-quality honey. The project also trained the hive makers how to make modern hives; the beekeepers choose the hives they want.

Beekeepers These producers have between 10 and 200 hives each, with which they produce honey and other products such as propolis and pollen (used as medicines) and beeswax. By December 2008, around 2,000 farmers, including 500 women, were producing honey as a business venture. The peak production season is between January and April, during the dry season following the October rains.

Box 6.1 The K-Rep Group

Founded in 1984, K-Rep began as a technical assistance and financing agency for non-government organizations. It has evolved into an umbrella organization with various units dealing with microfinance in Kenya (Figure 6.6).

- **K-Rep Development Agency** (KDA) This is the group's research and development arm. It identifies, develops, tests and refines financial products for marginalized groups in Kenya.
- **K-Rep Bank** This microfinance bank targets mainly small and medium enterprises. It is the second biggest microfinance bank in Kenya, and the sixth largest bank overall in the country, with 61,300 clients, a loan portfolio of nearly KSh 6 billion ($75 million) and a savings portfolio of KSh 7 billion ($88 million).
- **K-Rep Advisory Services** A consultancy company in microfinance and business development services.
- **K-Rep Fedha Services** This company manages village financial services associations. In December 2008, it managed 56 such associations, with a total membership of over 100,000 and an outstanding loan portfolio of KSh 300 million ($3.75 million). These loans are financed by the members' share capital and their savings (totalling about KSh 275 million ($3.44 million), and loans of about KSh 25 million ($310,000) from KDA and other sources.
- **Juhudi Kilimo** The latest entrant to the group, this company was registered in July 2009. It has evolved from what was KDA's micro-leasing project. It will primarily target the promotion and development of agribusiness value chains in Kenya.

K-Rep is a non-profit organization. Projects that begin making profits sustainably are spun off to form independent, for-profit companies. The honey value chain financing project was hosted by the micro-leasing project, which is now a separate company under the K-Rep umbrella. KDA aims to leave the honey project with its sister companies, K-Rep Fedha and K-Rep Bank.

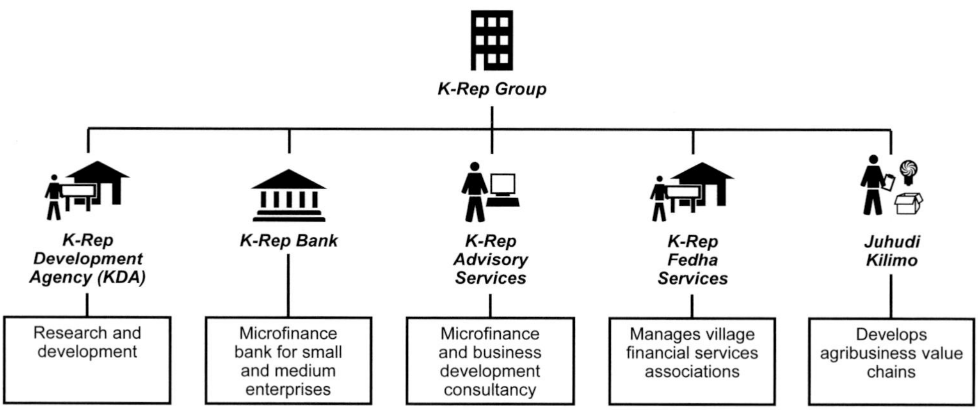

Figure 6.6 The K-Rep Group

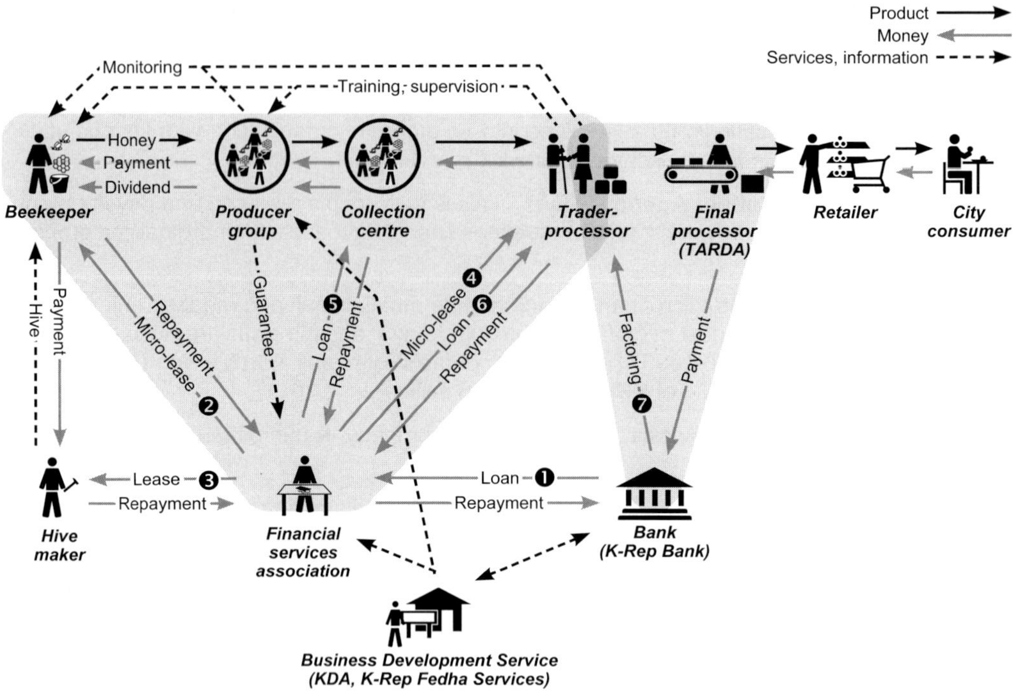

Figure 6.7 The honey value chain since 2007

Producer groups The beekeepers were organized in groups of between 20 and 35. These groups collect the crude honey and sell it to the collection centres. Because the honey is bulked, it fetches a better price than before.

Collection centres The collection centres are owned by the producers through shareholding. These buy honey, remove foreign matter, grade and bulk it, and store it before selling it to traders. By December 2008, there were 10 collection centres, each serving around 20 producer groups.

Traders The project trained the traders who used to buy low-quality honey from the beekeepers. These traders now buy bulk honey from the collection centres, and press and centrifuge it to separate the honey from the honeycomb. They then deliver the honey to the final processor. A majority of the traders are women; many of them are widows. There are 50 traders in the four regions of the district served by the project.

Final processor Under a new national government, TARDA restored its honey processing activities. It buys the honey from the traders for final processing and packaging. It pays traders KSh 150 ($1.88) per kilogram of honey, compared to KSh 120 ($1.50) offered by others. One disadvantage is that TARDA cannot pay for the honey immediately upon delivery – making an alternative system of payments necessary.

Retailers TARDA packs the honey and sells it to local retailers in Kitui and nearby towns, as well as to food processors, herbalists and pharmaceutical companies. Some companies also export the honey.

Consumers Consumers include households, hospitals and hotels.

Serving the chain

Financial services associations K-Rep Fedha set up village-based cooperatives known as financial services associations throughout the district to offer simple credit, savings and money transfer services to their members. These are user-owned and user-financed institutions, managed by K-Rep Fedha Services. Each association has about 2,000 members each: farmers, beekeepers and other local people, who own shares in it. The associations tailor loans to their members' needs; for example, there are loan products for cotton farmers, school fees, and so on. To qualify for a loan, borrowers must be a member of a group, and the group must give them a recommendation. They do not need hard collateral, but the group as a whole takes responsibility for repaying the loan. The financial services associations may borrow from K-Rep Bank or other lenders to cover fund shortages (❶ in Figure 6.7).

K-Rep Group Various units of the K-Rep Group (Box 6.1) provide services to the chain: KDA conducted studies and redesigned the chain; K-Rep Fedha Services manages the village financial services associations; and the K-Rep Bank provides financial services to various actors in the chain.

Memoranda of understanding

The K-Rep Group developed several financial services to help the honey value chain function more smoothly. Each type of service is designed for a particular stage in the chain, from beekeeper to final processor. They are based on close cooperation between the chain actors. In fact, that cooperation is so important that various memoranda of understanding have been signed between different parties to ensure it works smoothly. These are not contractual but are based on trust and goodwill. They ensure continuity in the chain because they are between organizations rather than the individuals who work for them. They are the first line of collateral for the financial organizations that provide loans and leasing services.

There are eight different memoranda of understanding between various groups of actors and the financial service providers. They specify the roles of each signatory and the arrangements for buying, selling and financing honey and inputs, as well as services such as training and monitoring.

Micro-leasing

K-Rep and the financial services associations use "micro-leasing" to encourage various actors to invest in the honey chain. Honey production and processing require relatively large one-time investments in things like hives, bee colonies and processing equipment. Such purchases lend themselves to a leasing arrangement. Loans are more appropriate for situations where the borrower needs many different, smaller-value items (such as bags of seed or sacks of fertilizer) on a repeated basis (e.g., new seed every season).

The micro-leasing scheme works like this: K-Rep (or the village financial services association) buys an item (such as a hive or processing equipment), then allows the lessee to use it for a certain period, during which the lessee must pay for it in instalments. The item becomes the property of the lessee at the end of the period.

Advantages of micro-leasing

A lease has several advantages over a loan:

- The financial institution remains the owner of the item until the lessee has paid it off fully. That makes it easier for the bank to repossess the item if the lessee fails to keep up the payments.
- The financial institution has more control over how the money is used: it buys the item, so it can choose the supplier and make sure the item is good quality. With a loan, the financial institution has much less control over how the borrower uses the money once it has been handed over.
- The financial institution can negotiate with suppliers to provide items in bulk at a low price.
- The borrowers also benefit, since they are assured that the items are good quality and reasonably priced.

Micro-leasing services

K-Rep and the financial services associations offer leases to three types of chain actors: beekeepers, hive makers and honey traders.

Beekeepers A typical investment for a hive plus ancillary equipment is worth KSh 3,500 ($44). This represents an average of 40 months of income for these people, so almost unaffordable.

To get a lease, the producer must find at least two quotations from suppliers of hives and equipment, and submit these to the village financial services association. The association appraises the request, and chooses one supplier, to whom it pays 60% of the price. The supplier must deliver the items within 21 days, then receives the remaining 40%.

The beekeeper must pay off the lease in monthly instalments, over a 24-month period. There is a grace period of 6 months to allow time for bees to colonize the hive and the first honey to be harvested so the beekeeper has a steady cash inflow to pay for the instalments. Only after paying the final instalment does the beekeeper become the owner of the equipment.

To qualify for the micro-leasing arrangement, beekeepers must have received training in hive management. By December 2008, close to 2,000 farmers had done so. The beekeeper must also belong to a group, which guarantees the request for the lease. So there is a triangular relationship between the beekeeper, the group and the financial services association (❷ in Figure 6.7 and first column in Table 6.1).

For the duration of the lease, the trader and the group monitor the equipment to make sure it is being used properly. The lessees are required to report to the groups periodically on the status of their repayments, so enabling the group to take collective responsibility for the equipment. The financial services associations also train the groups on how to manage loans and cash flow, and on how to use the credit properly.

Hive makers The hive makers also need wood and other materials, and tools to build the hives. The financial services associations provide leases (or loans) for these items, along similar lines as for the beekeepers (❸ in Figure 6.7).

Traders The traders also need equipment to process the honey. The financial services associations lease this to them using a similar procedure to the micro-leases for beekeepers, though with much higher amounts because the equipment is more expensive (❹ in Figure 6.7 and Table 6.1) They have so far disbursed KSh 1 million ($12,500) in this way.

Other financial services

Collection centres: Short term loans for working capital

When they deliver their honey to the collection centre, the beekeepers naturally enough want to be paid straight away. But the collection centres do not have the cash to pay them until they sell the honey - and that may not be for several months. So the financial services associations loan them money so they can pay the producers immediately (❺ in Figure 6.7). The producer gets 80% of the price immediately, and the remaining 20% once the centre has sold the honey to a trader, minus a KSh 1 ($0.01) administration fee per kilogram. Any excess is paid out as dividends at the end of the year.

These loans are for a maximum of 6 months and are repaid once a month in equal instalments. The interest rate is 18% a year.

Table 6.1 Financial products in the honey value chain in Kenya

	Micro-leasing (❷❹ in Figure 6.7)	Loans (working capital) (❻ in Figure 6.7)	Factoring (❼ in Figure 6.7)
Product and financial flows			
Purpose	Enable producers and traders to buy hives and other equipment Borrowers pay for items in instalments; when fully paid, they own the item	Give trader working capital to buy honey	Enable end-processor to delay payment for honey until it is sold
Amount	For producers, first loans maximum KSh 17,500 ($220) For traders, up to KSh 150,000 ($1,875) first loans Financial services associations lend 5 times value of member's shares (compared with 4 times for other loans) Total KSh 1 million ($12,500) disbursed	First loan of KSh 150,000 ($1,875) Total KSh 1 million ($12,500) disbursed	Total KSh 2 million ($25,000) disbursed
Period	Maximum term = 24 months	Maximum term = 6 months	Maximum term = 3 months
Disbursement	60% of price paid to supplier on approval of lease Remaining 40% paid on delivery	80% first payment 20% later payment, less 20% service charge	After delivery of honey, trader takes invoice to bank Bank pays 80% immediately after invoice is authenticated
Repayment	Equal monthly instalments Amount depends on negotiations between financial services association and client, but aims at full recovery of lease amount	Equal monthly instalments	Buyer pays after 3 months Fees are deducted then
Interest rate	15% a year flat rate Interest rate is low to encourage producers	18% a year flat rate	20% a year flat rate
Transaction costs	1% lease application fee 1% lease insurance fund against death of client	1% application fee 1% insurance fund against death of client	1% loan application fee 1% loan insurance fund against death of client

Continued...

Table 6.1 (continued) Financial products in the honey value chain in Kenya

	Micro-leasing (❷❹)	Loans (working capital) (❻)	Factoring (❼)
Risk management			
Securitization	Peer pressure from group Leased item and household items pledged to group	Household items pledged	Due diligence on buyer including inspection of audited accounts, etc
Liability	Group members are individually and severally responsible	Trader	TARDA, trader
Information flows			
Information required to apply	Producer: Number of hives already owned (minimum 10) Number of hives colonized (minimum 5) For start-ups, proof of training in beekeeping Trader: Must have handled over 3 tons in previous season Maximum of 1.5 times size of working loan Memoranda of understanding	Number of groups and hives supplying the trader Projected honey harvests for season Previous season's cash flow	Agreement with TARDA Invoice from TARDA
Information required during season	Number of leased hives colonized Proof of visits by hive specialist to inspect for diseases and pests Projected harvests during season	Expected harvests	
Time lag between application and payment	7 days	7 days	Immediate payment on submission of invoice

Traders: Short-term loans for working capital and micro-leasing of equipment

The traders need money so they can buy the honey from the collection centres. K-Rep Bank and the financial services associations provide them with short-term loans of up to KSh 150,000 ($1,875) during the honey harvesting season – far more than the amount they could get as a first loan from other lenders (❻ in Figure 6.7 and second column in Table 6.1).

The loans and leasing arrangements are subject to memoranda of understanding between the trader, collection centre, the financial services association and K-Rep Bank, in which the trader agrees to buy all the honey that the collection centre has available, trains the producers and maintains their hives. The collection centre agrees to sell its honey to the trader, giving it an assured market and the trader an assured supply.

Final processor: Factoring to enable purchase of semi-processed honey

In value chains, especially towards the end, supermarkets and other buyers often pay for products months after receiving them. That leaves the sellers short of the cash they need to keep their operations going. The honey traders and TARDA, the final processor, suffer from this problem. They need constant cash so TARDA can buy the semi-processed honey from the traders, and the latter can continue to buy crude honey from the collection centres.

K-Rep Bank's solution is known as "factoring" or "invoice discounting" (see also the case on tea micro-factoring, page 125). When TARDA's refinery in Kitui town receives a delivery of honey, K-Rep Bank pays the traders 80% of the price on TARDA's behalf. Three months later, after TARDA has processed and sold the honey, it pays the bank the full amount, and the bank pays the traders the remaining 20%, minus a fee equivalent to 20% a year flat rate. This agreement between K-Rep, TARDA and the traders is the second triangle in the chain (❼ in Figure 6.7 and third column in Table 6.1).

This arrangement ensures there is enough liquidity in the chain for it to work smoothly. The traders get their money on time, so can buy more honey in turn from the collection centres. This also prevents honey going to waste during the harvest season because of actors further down the chain not having the immediate cash to buy it.

Risk analysis

Because Kitui is a dry district, it has very limited business opportunities and high risks, and is not an attractive location for financers. Extended periods of donor programmes have left residents expecting handouts rather than loans they have to repay.

Table 6.2 summarizes the risks involved in financing the honey value chain.

Benefits

These innovations have resulted in benefits for various actors in the chain.

Beekeepers The leasing arrangements have enabled even the most disadvantaged producers to acquire new hives and equipment, and to upgrade their existing hives. They have been trained on beekeeping, and now have savings and money transfer services. They have increased both the quantity and quality of the honey they produce: in 2006, they harvested 2,500 tons of honey, while in 2007,

Table 6.2 Risk analysis for the honey chain in Kenya

	Source of risk	Risk level	Mitigation strategy
Production	Climate: bees leave in prolonged drought	●●●	Help producers diversify to poultry production, drought-resistant crops, etc.
Price	Prices high when production is low Producers side-sell if they can get better prices	●●○	TARDA survey at start of season to determine market prices TARDA pays higher than going prices
	Price fluctuations	●●○	Relation with TARDA helps stabilize prices TARDA pays higher than going price
Market	Disruption to continuity of chain	●●○	Memoranda of understanding between chain actors and financial service providers
	Single buyer: whole chain is threatened if TARDA cannot take product	●●○	Seek additional buyers
Default	Funds used for other purposes	●●●	Select traders who have agreement with TARDA Train clients before disbursement
	Low honey quality Borrowers expect handouts	●●○	TARDA continuously trains traders on controlling quality TARDA provides standard containers to conserve quality Traders have to take back inferior honey

Risk levels: ●●● High ●●○ Medium ●○○ Low

they produced 4,000 tons. In 2009, they hope to increase their output further. The improved quality is reflected in higher prices: in December 2008 a kilogram of honey fetched KSh 89 ($1.10), compared to KSh 60 ($0.75) a year earlier. This has made honey an important source of income for many people in a drought year.

Farmer groups and collection centres The farmer groups and collection centres have been able to get loans that let them pay producers immediately for the honey they deliver. The collection centres paid producers a total of KSh 720,000 ($9,000) in 2008: KSh 480,000 ($6,000) more than they would have got under the old system.

Traders Access to credit and the ability to lease equipment mean that traders can buy and process more honey. Improved quality and an assured market give them higher prices: their prices rose by 25%.

TARDA Before, a lack of financing forced TARDA to reduce its honey operations drastically. Factoring gives TARDA and the traders enough liquidity to continue buying honey throughout the year. From a low of just one ton, in 2008, TARDA processed 3,000 tons, and in 2009 it expects to handle 10,000 tons.

Financial institutions The financial services associations have increased their membership and loan portfolio by over 20%. Some 2,000 farmers have joined the associations, borrowing a total of KSh 3 million ($37,500) in the first year.

K-Rep Bank lent KSh 5 million ($62,500) in the first year to traders, a figure that is expected to rise to KSh 30 million ($375,000) as factoring becomes more common. The bank has eagerly embraced this model because though of medium risk, it enables the bank to do brisk business in this arid and semi-arid district.

Chain as a whole The package of innovative financial services provided to different actors has revived the honey industry in Kitui. The quality and output of honey have risen. There is a lot more liquidity in the value chain, with each actor now able to handle more honey because the financial constraints have been removed. Prices throughout the chain have risen, and the chain is now more organized and streamlined. The level of trust among actors has risen, cemented by a network of memoranda of understanding.

Factoring is still in a pilot stage, and lessons on how to apply it are still being learned.

One of the objectives of KDA's intervention was to stimulate interest in the value chain among other financial institutions. One of these, Equity Bank, is developing a loan product targeting beekeepers.

Challenges

Producers

- With better honey prices, the theft of hives has risen significantly. Farmers are urged to move their hives nearer to their homes so they can keep an eye on them.
- Beekeeping is traditionally a male affair, and few women are yet involved.
- Prolonged drought (sometimes lasting up to 7 years) leads to dismal honey harvests.
- Tree felling to make charcoal continues unabated. This depletes the vegetation vital for bees to make honey.

Financial service providers

- Financial services associations allow producers up to 6 months before they have to start repaying their lease. These long grace periods pose cash flow challenges, because the associations reply on their members' share capital for loan funds. As a result, they have had to look for other sources of money to loan out.

Lessons

- The project has shown that commodity value chains require different types of interventions, including business development and financial services.
- There is huge growth potential for chain actors and financiers in commodity value chains. However, value chains may derail at the pre-production stage, so a thorough study of this stage is important. For example, the hive makers in Kitui could not make the hives the farmers needed because they had no capital. And without transport, the hive specialists could not reach the farmers. The project has financed bicycles and motorbikes to enable them to provide their services to distant farmers.
- Appropriate financial interventions can make a major contribution to reviving and strengthening a value chain.
- TARDA has not been a reliable partner in the past, and overdependence on one government agency has to be avoided.
- As with other programmes that become sustainable, KDA aims to withdraw from its involvement in the chain by the end of 2009. K-Rep Bank and K-Rep Fedha Services will continue to be involved in the chain through the financial services associations.

More information

Paul Maundu Mwilu, *K-Rep Development Agency, maundumwilu@yahoo.com*

www.k-rep.org

Financing the soybean value chain in Ethiopia

"IT DOES NOT MATTER if we have no grazing land to rear cows", says Abba Sherab Abba Warri. "After all, soybean is our cow that never stops giving milk."

Abba Sherab is one of many farmers in Jimma zone, in Oromia state in southwestern Ethiopia, who have recently taken up growing soybeans. He sells the soybeans to the marketing organization he belongs to, which takes them to a factory nearby which produces soy milk, a nutritious substitute for cow's milk. The soy milk is proving popular in the city of Jimma, the capital of the zone. Cow's milk is not readily available in the area, and local Orthodox Christians abstain from animal products for about 200 days a year.

The farmers in Jimma zone face many problems. The soil is poor, and many farmers plant maize, the staple food, year after year. They have no money to buy inputs such as seed and fertilizer, and they lack both a market for their surplus produce and the skills to sell it. Borrowing money costs a lot: moneylenders charge 120% interest a year. Farmers who take up such loans are sometimes forced to sell their produce at low prices to pay off their debt.

Some farmers own a few cattle, but this is a densely populated area and grazing land is scarce. The cattle are of low quality, and drought and diseases mean that most cows produce less than a litre of milk a day.

A plant that gives milk

What alternative crops could farmers grow? Facilitator for Change Ethiopia (FCE), a national NGO working on community development (Box 6.2), discussed various options with local farmers, research centres and government agencies, and came up with soybean. This legume is well-adapted to the local soils and climate, and improves the soil fertility by fixing nitrogen. The beans can be used for food and feed as well as processed into various products, including milk. Market studies showed

Figure 6.8 Location of Jimma, Ethiopia

it had a promising market in the city of Jimma, as well as among food processing plants in Addis Ababa.

FCE and Jimma Agricultural Research Centre have jointly promoted soybean in Jimma zone since 2006. They have put a lot of effort into popularizing the crop among farmers: each year a soybean field day is held, farmers visit model farms, and exhibitions are staged where local people can learn about foods made from soybean. As farmers gradually adopted and began to consume soybean, the need to find a bigger market for the beans became clear.

At the same time, women's groups in Jimma city were looking for ways to earn money. Familiar with the shortage of cow's milk, they decided to make and sell soy milk. They approached Harbu Micro Finance Institution, an affiliate of FCE, to discuss this idea. FCE decided to link the women with the existing farmers' groups to form a new value chain.

The soybean value chain was established in the city of Jimma and three adjacent rural districts, Tiro Afeta, Kersa and Omo Nada. It links smallholder producers, farmers' marketing organizations, the women's groups (grouped together into "cluster-level women's associations"), retailers and consumers.

Chain actors

Smallholder farmers A typical smallholder farmer owns 1.5–3.5 hectares of land and grows soybean on about one-quarter of this area. Most farmers in the

Box 6.2 Organizations supporting the soybean value chain

Facilitator for Change Ethiopia (FCE)

Established in 1997, FCE is a national NGO that helps disadvantaged communities overcome poverty. It focuses on enabling them to improve their livelihoods using their own resources, skills and experiences, with minimum external support. FCE currently has seven projects in Oromia and Amhara states in Ethiopia, three of which promote farmer marketing organizations.

FCE is heavily involved in value chain promotion, especially on soybeans. It has organized about 5,000 farmers into 64 farmer marketing organizations in these two states. For the soybean value chain in Jimma, FCE is supported by SOS-Faim, a Belgian-based NGO.

Harbu Micro Finance Institution

This was established in 2005 and is affiliated to FCE. It aims to boost agricultural productivity and agricultural marketing by supporting value chain development and access to financial services. It has capital of 12.4 million birr ($1.24 million). Harbu is registered with the National Bank of Ethiopia to mobilize savings. It receives support from Terrafina Microfinance (a Dutch microfinance promoter) and SOS-Faim.

More information: www.fce-eth.org

area earn around 4,500 to 5,000 birr ($450–500) a year. They sow soybean between the end of May and mid-June, and harvest them in December. The farm families consume about 25% of the harvest themselves, or save it as seed for the next season; they sell the rest to farmers' marketing organizations. The average yield is 1.5 tons a hectare.

Farmers' marketing organizations Many of the farmers in Jimma zone are organized into neighbourhood groupings known as *ollas* consisting of 20–25 farmers each. Several *ollas* are grouped into voluntary associations known as farmers' marketing organizations. Throughout Ethiopia, FCE has helped establish 64 of these marketing organizations, with a total of over 4,000 farmer members. There are 30 marketing organizations in Jimma zone, with over 2,000 members and a working capital of birr 750,000 ($75,000). FCE helped them get organized, built their capacity, and provided start-up capital so they could begin trading. They have gained a lot of experience trading in grain and business activities.

Five of the marketing organizations in Jimma zone, with a total of about 240 members, are engaged in the soybean milk value chain. Between them, they have around birr 150,000 ($15,000) in working capital. They buy soybeans from their members, clean them and pack them ready for sale. Each organization has a warehouse where it stores the bulked soybeans. The organization's managers negotiate the sale of the beans.

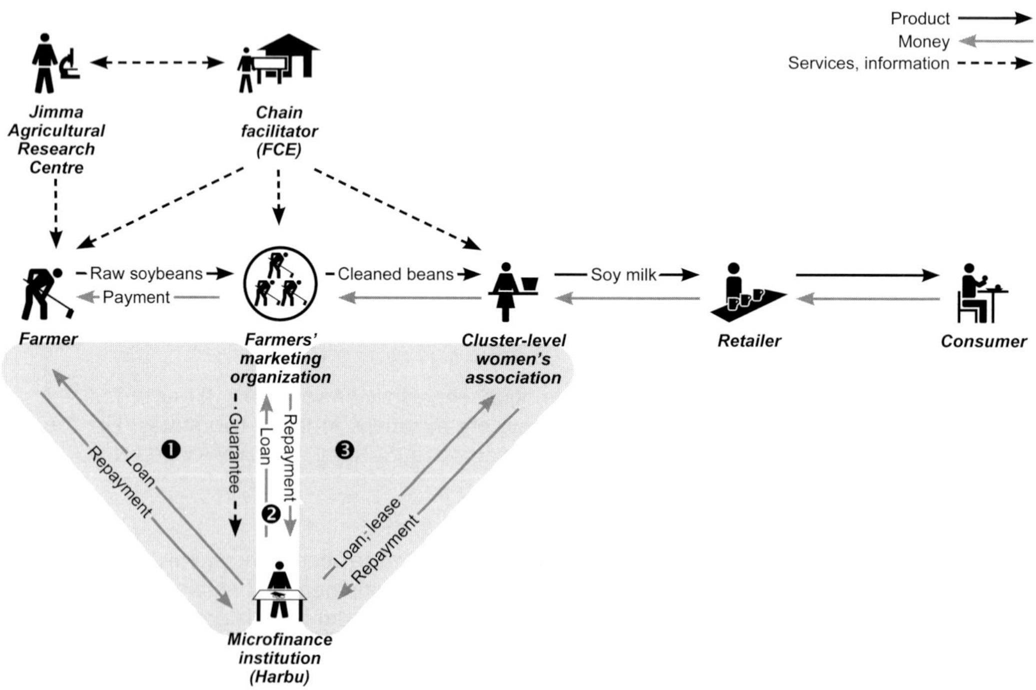

Figure 6.9 The soybean value chain

Figure 6.10 Soybean is a new crop in the Jimma area

Cluster-level women's associations Since 2000, FCE has helped establish 131 informal self-help groups of poor women in Jimma zone, each with 15–20 members. These groups in turn make up 14 cluster-level women's associations, which are engaged in various types of small enterprises, services and manufacturing. The associations aim to give their members access to finance, business skills and other resources. Involvement in these enterprises makes membership of an association attractive: the average annual income of members had risen from birr 1,300 ($130) before joining the group, to 3,600 by December 2008.

One of these associations, with 164 women members, is involved in the soybean milk value chain. It buys all the soybeans sold by the five farmer marketing organizations, plus soybeans from other farmers' marketing organizations in the area. It soaks the beans to remove the hulls, grinds them up in water, and filters the resulting liquid to make soy milk. This work is all done in a factory, managed by representatives elected from among the association. Some association members work in the factory (they earn a wage for their work); the association hires professionals to run the equipment.

One kilogram of raw soybeans produces 7.5 litres of soy milk. Members consume 5% of the resulting milk at home. They fill the remainder into half- or one-litre bottles or 5-litre cans, and sell it to retailers and other customers in Jimma. The association also makes and sells tofu and animal feed from soybeans.

Retailers The retailers are individual entrepreneurs around Jimma city who have relationships with the associations. They sell the soy milk to consumers.

Consumers The soy milk is aimed at consumers who readily accept new ideas: university students (who account for 10% of the market) and the middle class, in the hope that the product will spread to other groups. Many consumers prefer soy milk to cow's milk because it is good quality, more nutritious and costs less than cow's milk (much of which is diluted). Other consumers include children, cafeterias, hospitals, and other urban residents. The milk does not yet reach buyers outside the Jimma area.

FCE and the Jimma Agricultural Research Centre have jointly promoted soybean in Jimma zone since 2006. Continuous training on food preparation has been given to the wives of the soybean producers, so soy-based food have become included in many families' daily meals. A Dutch organization, SharePeople, supported FCE and Harbu in testing the product in universities and cafeterias to check for people's preferences. The comments and feedback from this market research were encouraging.

Figure 6.11 The women's association processes the beans into soy milk

Establishing the chain

FCE has played a lead role in building and strengthening the chain and its main actors. It organized both the farmers' marketing organizations and the women's association, and built their capacity through training and experience sharing. It guided them on how to manage internal loans and funds from outside. It had worked with these groups before the soybean value chain was created, so had already developed a trusting relationship with them.

With support from Jimma Agricultural Research Centre and rural development specialists, FCE raised understanding among the farmers on soybean production. To help farmers find a reliable market outlet, FCE provided the marketing organizations with seed capital, material and technical support.

- **Seed capital** To encourage the marketing organizations to build their working capital and develop a saving habit, FCE provides them with matching finance on a 50:50 basis: for example, if a marketing organization has raised 20,000 birr, FCE provides it with another 20,000 birr. This financial support may last for only 3 consecutive years after a marketing organization is established. The upper limit of financial support is usually 40,000 birr.
- **Material support** When the marketing organizations have mobilized locally available resources such as land, wood, and labour (to build a warehouse, for example), FCE provides purchased inputs such as nails, metal sheets, cement and payment for a contractor.
- **Technical and capacity building support** FCE provides professional advice, trainings, experience sharing etc. Like the other types of support, there is a farmer contribution here too. For example, for training courses, the marketing organizations cover the transport costs while FCE covers per diems and professional fees.

FCE also linked the women's association to the farmer organizations, input suppliers and customers, as well as to Harbu. It helped test consumer preferences for things like packing materials for soy milk, the size of bottles, flavour, place of delivery, etc.

The Ethiopian government controls many sectors of the economy, so is an important influence on the chain. It supports the sale of inputs to smallholder farmers and provides extension services and technical support. Through line bureaus such as the rural development office and the cooperative promotion commission, it also provides services such as farm management, licensing of cooperatives, auditing and quality control.

Financing the chain

Harbu provides a variety of financial services to the value chain: agricultural loans, working capital and lease-financing loans.

Agricultural loans for farmers The soybean growers need money to buy inputs such as seed, fertilizer and implements. Harbu is the only financial institution in the area that gives small loans to farmers, as other financers find agriculture too risky and demand hard collateral from the farmers. To qualify for a loan, a farmer has to be a member of a farmers' marketing organization, and the organization must have been operational for more than a year. If an individual borrower fails to repay a loan, the *olla* group and the farmers' marketing organization are responsible for repaying it. This financial arrangement is shown as the first triangle in the chain (❶ in Figure 6.9).

Working capital loans for farmers' marketing organizations Many of the farmers' marketing organizations have savings accounts with Harbu. But they need capital so they can buy beans from their members. Harbu loans them money based on their financial statements, warehouses and inventories. The government revenue department does not recognize the warehouses as assets that can be used as collateral, so they are considered as "soft" collateral. A contract between Harbu and the marketing organization covers details of interest rates, duration, repayment, etc. Five percent of the total loan must be deposited with Harbu as compulsory savings. All five marketing organizations in the soybean chain have received such loans from Harbu. This loan is shown as ❷ in Figure 6.9.

Loans and leases for women's association Harbu gives similar working capital loans to the women's association so it can buy soybeans from the marketing organizations and packaging materials from suppliers. In 2008, the association approached Harbu for a loan so it could buy a processing machine to make soy milk. This machine cost $25,000 – an investment that Harbu found too risky. So Harbu proposed an alternative: rather than providing a loan, it would lease the machine to the association over a 5 year period, charging 10% interest a year (the interest rate declines as the lease is paid, and the machine belongs to the association at the end of the period). If the association does not pay on time, Harbu can repossess the machine and sell it to somebody else. This arrangement is based on the long-term trust relationships between Harbu, the women's association and the marketing organizations. This is shown as the second triangle in the chain (❸ in Figure 6.9).

Table 6.3 gives further details of these financial products.

Risk analysis

Table 6.4 summarizes the risks in financing the soybean value chain. Most of the risks are mitigated through the triangular cooperation between FCE's development support, Harbu's financial services, and the intimate business relations between the farmers and the women groups. Without this institutional architecture, the risks and costs would be too high, and the soybean chain would not function properly.

Figure 6.12 Soybeans can be used to make a variety of foods as well as a good substitute for cow's milk

Benefits

Farmers Farmers in Jimma are no longer dependent on a single crop, maize. In soybeans, they can now grow a crop that they say raises their income by more than 7% and improves the fertility of their soil. Soybean is a nutritious food that can be consumed in various forms – not only milk, but also bread, cheese and other products. Soybean by-products are used for animal feed. In 2008 over 1,000 farmers started producing and selling soybean. They are now more able to grow crops for the market, and they can get loans to increase their productivity and incomes. Their ability to organize and promote sustainable local development has been increased.

Farmers' marketing organizations The marketing organizations have a reliable buyer for their product, and get higher prices than they would get if the farmers were to sell as individuals or on the open market. That has raised their profits by over 5% and improved their ability to serve their members. The relation between the women's association and the farmers' organization has improved the reputation of these farmer groups.

Women's association The women who work together in the association have also benefited. Their incomes have increased by more than 5%, and they have been able to scale up their business. The soybean chain greatly contributes to the social and economic empowerment of women in the area.

Harbu Harbu has increased its portfolio of products and its clientele base by 20%. It has lowered the risk of default by working with several actors in the chain rather than with individual players. By doing business with groups rather than individual borrowers, it lowers its risks and its transaction costs by at least 50%.

Chain as a whole Before this initiative, the soy milk value chain did not exist. Consumers in Jimma now have a substitute for cow's milk: a reliable supply of fresh, nutritious soy milk at an affordable price.

The linkages between rural areas and the town of Jimma have been strengthened. The chain has helped lower poverty, improve nutrition, and empower women and smallholder farmers. It has contributed to development of agro-processing enterprises and increased industrial development in the area.

Table 6.3 Financial products in the soybean value chain in Ethiopia

	Agricultural loan (❶ in Figure 6.9)	Working capital loans (❷ in Figure 6.9)	Lease financing (❸ in Figure 6.9)
Product and financial flows			
Purpose	Enable soybean producers to buy inputs	Enable farmers' marketing organizations and women's association to buy soybeans and supplies	Enable women's association to buy equipment
Amount	Average: $275 Max: $500 Min: $50 Total portfolio: $280,000	Average: $6,500 Max: $10,000 Min: $3,000 Total portfolio: $32,500	Avergage: 48,450 Max: $93,900 Min: $3,000 Total portfolio $96,900
Period	9 months	5 months (marketing organization) 1 year (women's association)	5 years
Disbursement	In May to farmer	In December to marketing organization	By Harbu to machinery owner
Repayment	Repaid to Harbu, payable all at one time	Repaid to Harbu, payable all at one time	Throughout the year
Interest rate	15% per year, payable all at one time	15% per year (marketing organization) 15% (women's association)	Initially 10%, declines as lease is paid
Transaction costs	Loan processing fee = 3%	3%	3%
Risk management			
Securitization	Peer pressure from group	Warehouse and inventory 5% of the total loan should be deposited as compulsory savings	Harbu retains ownership of equipment until payment is complete
Liability	Farmers' marketing organization, neighbourhood groups and individual producers	Marketing organization (not individual members)	Women's association (not individual members)

Continued...

Table 6.3 (continued) Financial products in the soybean value chain

	Agricultural loan (❶)	Working capital loans (❷)	Lease financing (❸)
Information flows			
Information required to apply	Group cohesion (number of conflicts, use of internal funds, % attendance at meetings, etc.) Purpose of loan Credit history of borrower Marketing organization must support application	Assessment of assets owned Balance sheet and income statement Audit reports Reputation (previous business track record, clarity and neatness of records, profitability, etc.)	Assessment of association's strength and experience in managing internal loans Purpose of loan Feasibility of the business to be undertaken
Information required during season	Crop assessment by Harbu Harbu attends fortnightly (initial, then monthly) meetings to monitor use of loan	Stock assessment every month by Harbu	Harbu attends monthly meetings, or refers to meeting minutes
Time lag between application and payment	2 weeks	1 month	1–2 months

Lessons and challenges

- There are limited funds for loans to meet the needs of different actors. Harbu will work with commercial banks to increase the amount of funds available for loans. By offering finance to several chain actors at the same time, it is possible to bring about a greater impact with the same level of inputs.
- To benefit from value chain initiatives and play a greater role in the chain, farmers and women need to be organized, linked to other groups, and have their skills upgraded. But the history of cooperatives in Ethiopia makes local people reluctant to get organized. Training, awareness raising and experience sharing from model cooperatives are necessary to overcome this barrier.
- Some chain actors have difficulty switching from competing with one another to cooperating. It is possible to change their mindsets by showing the benefits of the value chain. Creating access to markets and financial services improves production – it is estimated that some 7,500 farmers will be willing to grow soybean in the next 2–3 years because they see the benefits arising from the value chain.
- The lack of strong regulation to enforce contracts and quality standards hinders the standardization of services and products in the chain. There is need to lobby government to enforce its regulations so as to strengthen trust between the actors.

Table 6.4 Risk analysis for the soybean value chain in Ethiopia

		Source of risk	Risk level	Mitigation strategy
	Production	Crop failure due to poor rainfall	●○○	Buy soybeans from producers elsewhere
		Inadequate quality	●●○	Farmer organizations provide technical support to farmers Women associations pay premium for quality
		Inputs unavailable or late	●●○	Strengthen links with government agencies that provide inputs
		Failure of soy milk processing machine	●●○	Establish complementary enterprise: a flour mill
		Lack of management skills in women's groups	●●○	Train women and share experience
	Price	Rise in input prices	●●○	Provide loans to farmers to buy inputs
	Default	Beans not delivered because of low production or high home consumption	●●○	Build relations with farmers and raise awareness Women's associations pay competitive price and premiums
	Currency	Inflation	●●●	Gather, manage and use market information Women's association minimizes administrative costs and forecasts inflation

Risk levels: ●●● High ●●○ Medium ●○○ Low

The future

The future for the soy milk chain looks bright! A market and feasibility study by a private consultant and a group of Dutch entrepreneurs concluded that the soybean value chain in Jimma is an excellent example of social entrepreneurship on the basis of true economic principles. This study identified new retailers and consumers in Jimma city, and found that many institutions (universities, hospitals, supermarkets, cafeterias, hotels etc.) were interested in obtaining the milk. Expanding the chain would create further benefits for the actors involved.

More information

***Tesfaye Befekadu**, Harbu Microfinance Institution Share Company, harbumfi@ethionet.et, harbumfi@yahoo.com, tesfayebef@yahoo.com*

***Tarekegn Garomsa**, Facilitators for Change Ethiopia, fce@ethionet.et, taregarom@yahoo.com, website www.fce-eth.org*

Opening opportunities to small coffee producers in Nicaragua

COFFEE IS AN IMPORTANT crop in Nicaragua. Around 110,000 medium- and small-scale farmers grow coffee, and it accounts for 16% of the country's exports - about 93,000 tons of high-quality, unroasted coffee beans a year. Nicaraguan coffee is well recognized internationally for its quality and taste. Some 20% of the output is organic - which fetches a premium price in the market.

One of the major centres of coffee production is San Juan del Río Coco, in the highlands in the north of the country. Some 3,500 small- and medium-scale farmers around this small, highland town produce over 4,000 tons of coffee a year, 28% of which is organic.

Producing and processing coffee

The coffee production season in San Juan starts in March or April, when the farmers prune the bushes and plant new seedlings to rejuvenate part of their plantings. During the wet season, from May to September, the farmers have to control pests and diseases, weed the plantations and fertilize the bushes. The coffee bushes come into bloom (the first indication of how good the harvest will be), and then each blossom forms into a green berry, or "cherry". Pest and disease control continues during October and November. Finally in December to February, it is time to harvest the ripe berries, which are now bright red.

Figure 6.13 Location of San Juan del Río Coco, Nicaragua

The average coffee plantation size is 3.5 hectares, and produces about 0.5 tons per hectare of "green" coffee (beans ready for roasting). Farm families which have less than a couple of hectares can harvest their bushes themselves. Larger-scale farmers hire workers, or get their relatives to help. The harvest-

Box 6.3 UCPCO

UCPCO is a union of six village-level cooperatives of organic coffee producers in San Juan del Río Coco, with a total of some 400 members. The Union offers a number of services to its members:

- **Technical assistance** and training to improve the quality of production.
- **Production services** Collecting, dry processing, transportation, quality control and export.
- **Organic certification** for the farms, together with Bio Latina, an accredited certifier of agriculture and livestock production.
- **Working capital credit** The Union used to offer loans to cover members' working capital needs (see "Sources of finance" section below).
- **Community services** The Union runs community centres, programmes on sport, gender and women's health, and training activities. It also offers scholarships and school supplies for children, as well as medical support for special cases.

Various NGOs and research institutes (SNV, Pasolac and CATIE) support the Union's institutional capacity building and technical assistance activities, as well as providing finance to support loans for harvesting and processing.

In 2007, the Union was named Innovative Exporter by the Nicaraguan Producers' and Exporters' Association of Non-Traditional Products. In 2008, it received a national exporters' prize from the Ministry of Promotion, Industry and Commerce.

ers pick the ripe cherries, take them for weighing, then tip them into a concrete basin full of water. The good beans sink, while unripe beans, leaves and twigs float. The cherries then go into the farmer's depulping machine, which removes the skin and flesh, leaving pale-coloured beans, called "parchment" coffee. (If the farmer does not own a depulper, he or she may borrow one from a neighbour.)

The farmer then takes the parchment coffee to UCPCO (Box 6.3), which weighs the beans and pays the farmers 50% of the agreed price. It then sun-dries the beans, classifies them into quality grades, and puts them into bags. It transports the bags to the port in a closed lorry to make sure they are not contaminated. At the port, an agent arranges for the coffee to be shipped to its destination. After that, the coffee must be roasted, blended, perhaps ground, then packaged before it appears on retailers' shelves.

Sources of finance

Farmers need working capital during the season to pay for inputs and hire workers, as well as investment capital to improve their processing facilities so they can produce the quality required by the demanding organic market. They could only cover part of these financing needs, using three sources: UCPCO, private traders, and the Fondo de Desarrollo Local (FDL, see Box 6.4).

Box 6.4 Fondo de Desarrollo Local

Fondo de Desarrollo Local (FDL) is a microfinance institution focusing on rural areas. FDL offers credit to poor families, and provides technical assistance and training through an alliance with Nitlapán, a research institute at the Central American University in Managua. It is the largest non-regulated microfinance institute in the country (its non-regulated status means it is not allowed to accept deposits from clients) and the third-largest overall. Its loan portfolio is $70 million, and it serves 83,000 clients through a network of 36 agencies. Over 60% of the loan portfolio is dedicated to agricultural and livestock production, serving 40,000 producers.

With its commitment to development, FDL has designed financial products to promote investments in rural infrastructure, equipment, improved production, etc. It dedicates 15% of its loan portfolio to finance activities with special conditions, such as lower interest rates and longer repayment terms. It complements these financial services with technical assistance to improve production in an environmentally friendly way.

FDL is a leader in financing the agricultural sector in Nicaragua and was named the best microfinance institution in Latin America by the Inter-American Development Bank in 2005.

UCPCO loans to farmers The Union used to provide loans to its farmer members at the beginning of the season, so these had working capital to buy inputs and hire labour (❶ in Figure 6.14). When the farmers delivered the parchment coffee, the Union would pay them half of the agreed price immediately. It paid the balance, minus the loan amount and interest, after the coffee had been sold to international buyers.

At the beginning of each production cycle, the Union negotiated prices and signed contracts with buyers. The buyers paid only after receiving the coffee, usually after 3 months. They did not finance the coffee production or in-country processing.

Loans from private traders Farmers who needed cash to cover their production costs could also borrow from coffee trading houses, or their intermediaries. These lenders would charge much higher rates of interest for this finance than the Union - effectively 30–100%. The farmers would be obliged to sell their coffee to the traders rather than to the Union (❷ in Figure 6.14).

FDL loans to individual farmers The microfinance institution provided individual farmers with short-term loans for 4–12 months (it was not working with UCPCO at this stage). The individual borrowers were responsible for repayment (❸ in Figure 6.14).

Problems in financing the chain

This system faced several problems.

- **Lack of liquidity** The Union had to use a significant amount of its own funds to provide working capital to the farmers. But it sometimes did not

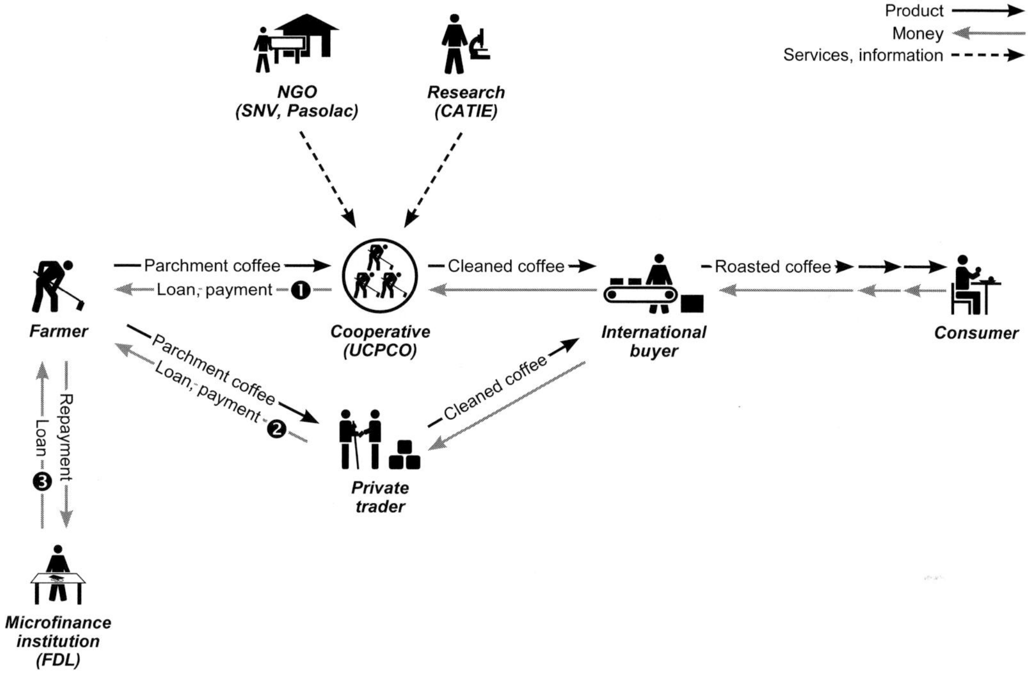

Figure 6.14 The original coffee value chain

have enough cash available, so had to reduce or delay its disbursements to the farmers and could not buy all the coffee the farmers had to sell.

- **Dependency on a few buyers** With limited capacity to buy coffee and attract new members, the Union had only a limited amount of coffee to sell, so was restricted to only a few buyers.
- **Lack of investment** Low turnover and lack of capital meant that the Union had few funds to invest in the chain. Without access to investment finance, the producers could not afford to buy new equipment. Lacking adequate processing facilities, the farmers would deliver lower quality beans.
- **Lack of technical assistance** Many of FDL's farmer clients were not members of UCPCO, so had limited access to technical assistance and usually received lower prices for their coffee. That increased FDL's risks in dealing with them.
- **No benefit from organic coffee** The private traders were not part of the organic value chain, so did not pay the premium price due to organic coffee. A potentially high-value product either ended up as regular coffee, or was sold as organic without the producer benefiting.
- **Overconcentration on coffee** FDL's portfolio was highly concentrated on coffee in San Juan del Río Coco, so price fluctuations became a major risk. The producers specialized in coffee, so had few other sources of income. That heightened the risk of farmers defaulting on their loans.

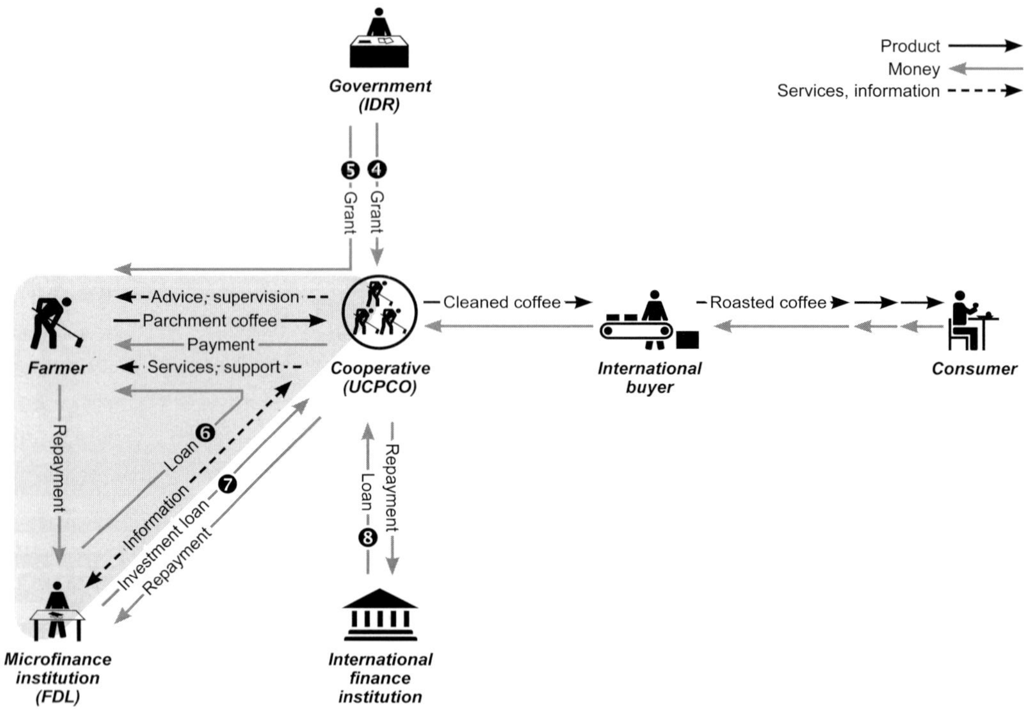

Figure 6.15 IDR and FDL support the coffee chain

Improving the chain

To revitalize the coffee chain, the chain actors needed capital to invest in technology, improve the quality of the coffee, and increase their scale of operation. UCPCO contacted FDL and the Instituto de Desarrollo Rural (IDR), a government rural development agency to discuss how to do this. The main problem the Union wanted to solve was the lack of depulping machines, concrete basins and other processing equipment. It also needed funding to improve its support for its members.

IDR support

UCPCO negotiated two lines of support from IDR:

Technical assistance grant to UCPCO This 3-year, $95,000 grant covered the cost of a technical assistance programme to improve production techniques and product quality, and to reduce water pollution from the farmers' depulping machines (❹ in Figure 6.15). Training of small-scale producers was a major component of this activity.

Investment grants for farmers IDR provided capital to farmers so they could upgrade their processing equipment. The improved equipment cost $2,500 per farmer. IDR was willing to provide 70% of the total. IDR pays this to the Union, which is in charge of buying equipment and materials and distributing them to the farmers (❺ in Figure 6.15). The farmer has to find the remaining 30% of the total.

FDL support

Investment loans for farmers It was difficult for the small-scale producers to raise even the remaining 30%. So Union approached FDL to develop a microfinance package that would give them the money they needed – an average of about $750 per farmer. FDL agreed to finance a 4-year loan at an annual interest rate of 14%. This money is disbursed to the farmers via the Union (❻ in Figure 6.15). Details are given in Table 6.5.

To qualify for a loan, a farmer has to have a plan for how to use the money to improve the wet-processing facility. The farmer has to be approved by the Union, agree to deliver his coffee to the Union, allow inspection visits by FDL credit officers, and provide information about the previous production cycle. Farmers borrowing less than $1,500 (which includes most borrowers) are organized into "solidarity groups" of 3–7 farmers: if one member defaults, the other members of the group have to pay the balance of the loan. Farmers borrowing more than $1,500 have to provide hard collateral in the form of livestock, equipment, vehicles or land. Union technicians visit the farms regularly and report on progress and problems to FDL.

After harvesting, the farmers deliver their coffee to the Union. The Union pays 50% of the price to the farmers immediately, then deducts an instalment of the outstanding loan, plus interest, from the balance after it has received payment from the buyer. It then pays the remaining amount to the farmer. If the balance is insufficient to cover the amount due, the Union advances the difference and charges the farmer the next season.

Such loans are denominated in córdobas, the Nicaraguan currency. The country's central bank maintains a "crawling peg" devaluation system, and loans for the whole financial system are indexed to the US dollar. FDL mitigates the currency risk by including a clause in its loan contracts to maintain their value against the dollar, based on an annual devaluation rate of 5%.

Working capital loans for farmers FDL has expanded its working capital loans to Union members, thereby freeing UCPCO from the need to lock its capital in cash payments to the farmers while waiting for the buyers to pay after 3 months only. With FDL taking over the financial services to the farmers, UCPCO is able to focus on its core business: technical assistance to the farmers and the buying and selling of coffee. The FDL loans are for up to 12 months and interest rates are higher – from 18 to 21% per year. Borrowers are not required to specify how they will use the money, so no investment plan is required. Other conditions are similar to FDL's investment loans, so this is included in arrow ❻ in Figure 6.15. Details are also given in Table 6.5.

Figure 6.16 The value chain's financing is based on a three-way partnership between FDL, farmers and UCPCO

Investment loans for UCPCO FDL is prepared to offers loans of up to $50,000 to the Union to improve infrastructure, buy vehicles, and pay for other major capital expenses. The Union has so far borrowed only up to $20,000 at a time. The Union has to provide hard collateral in the form of a mortgage of its land or buildings. Such loans have a 4- or 5-year term and must be repaid in annual instalments. Because the Union now has cash in hand, it has been able to repay these loans early. This arrangement is shown as ❼ in Figure 6.15.

Working capital loans for UCPCO FDL is also willing to provide working capital loans to the Union so it can buy coffee during the harvest season from members or from non-members. The Union has not yet needed to take advantage of this facility as it now has cheaper alternative sources of finance. Three development financial institutions, Root Capital, Rabobank Foundation and Shared Interests, have started to give loans to the Union for working capital and investments (❽ in Figure 6.15). This is possible because the value chain has grown and the Union has developed a good reputation in financial circles.

Table 6.5 FDL's loans to farmers (❻ in Figure 6.15)

	Investment loan	Working capital loan
Product and financial flows		
Purpose	Finance investment capital needs of UCPCO members for equipment, new coffee plantings, and rejuvenating old plantations	Finance working capital needs of UCPCO members Covers food expenditure of producers who use family labour
Amount	Up to $10,000 Average less than $1,500 Total portfolio: $92,000	Up to $5,000 Average less than $1,000 Total portfolio: $346,000
Period	2–4 years, depending on borrower's ability to pay	4–12 months, depending on loan date and nature of borrower
Disbursement	Individual borrower signs credit contract. Disbursement via Union Max 3 equal instalments	Individual borrower signs credit contract. Disbursement via Union Lump sump
Repayment	Annual payments after harvest (Jan–Apr)	At loan maturity (Jan–Apr)
Interest rate	14% + 1% administrative costs	18–21%
Risk management		
Securitization	Up to $1,500: no collateral, but solidarity guarantor required $1,500–10,000: collateral such as livestock, equipment, harvest or land	Up to $1,500: no collateral, but solidarity guarantor required $1,500–10,000: collateral such as livestock, equipment, harvest or property
Liability	Shared between UCPCO and FDL Borrower provides guarantee	Shared between UCPCO and FDL Borrower provides guarantee
Information flows		
Information required to apply	UCPCO membership and consent Farm development plan Identity document Farm visit by FDL Info on previous year's production Coffee to be delivered to UCPCO Signed contract For loans over $1,500: • Membership of solidarity group or individual credit • Solidarity guarantor, collateral	UCPCO membership and consent Identity document Visit to farm by FDL staff Info on previous year's production Coffee to be delivered to UCPCO Signed contract For loans over $1,500: • Membership of solidarity group or individual credit • Solidarity guarantor, collateral
Information required during season	UCPCO provides information on borrower's performance before FDL makes subsequent disbursements This information is collected by UCPCO technical assistant	Periodic information about the borrower's performance collected by UCPCO technical assistant
Time lag between application and payment	Maximum 8 days	Maximum 8 days

Mitigating risk

It was important for FDL to reduce its risk exposure as it had been allocating investment loans to farmers without collateral. FDL had already explored various alliances to support coffee value chains, but these had failed, or benefited the trading house rather than the producers (FDL's target group for its development activities). So FDL was searching for new models. The new arrangement is based on a strong triangular alliance between the farmers, the Union and FDL that is central to reducing risk.

The following points help mitigate FDL's risks:

- Technical assistance is available to improve the quality of the output.
- Farmers have enough cash to prepare their plantations properly and control quality.
- The farmers must be members of the Union, which has the ability to monitor and coordinate production, and which shares information with FDL.
- The Union retains part of the payment for the coffee the farmers deliver, and repays this to FDL.
- FDL takes a conservative approach in its credit analysis. It uses conventional coffee prices, not organic, to calculate the farmers' ability to repay the loans.

Table 6.6 summarizes FDL's risks in financing the value chain.

Benefits

Farmers The farmers now have access to investment funds with longer repayment terms and lower rates of interest. They are confident that they will be able to get a loan from FDL when they need it.

Coffee producers increased their income by an average of 7.7% in 2006–7, and again in 2007–8. This was mostly due to better quality and prices, rather than an increase in productivity.

UCPCO The Union's partnership with FDL has increased its liquidity and purchasing power: it can buy more coffee during the harvesting season. It now has access to investment funds from FDL and international development finance organizations.

Freed from the distraction of having to fund and manage its own microfinance programme, the Union can specialize on its core business – providing services to its members, assuring quality, and marketing the coffee. Its exports have increased from under 100 tons in 2005 to over 500 tons in 2008 (Figure 6.17).

The Union has grown from 117 members in 2005 to 400 in 2008. It has attracted new members, including small-scale producers who were already clients of FDL. It has expanded its range of international buyers and has built strong relationships with them. It is now in a position to reject new buyers, so enjoys a stronger

Table 6.6 Risk analysis for coffee in Nicaragua

	Source of risk	Risk level	Mitigation strategy
Production	Climate	●●○	Visits by UCPCO staff to farmer at least three times during production cycle and reports to FDL Check forecasts and expected yields Adjust terms of loans accordingly (e.g., extend repayment period if harvest is late)
	Disease	●●○	UCPCO staff advise farmer how to manage disease
	Quality	●○○	Almost all producers have wet process equipment Others use neighbour's equipment
	Crop cycles	●○○	Coffee harvests follow a cycle: high one year, and low the next Loans amounts adjusted accordingly
Price	Market price fluctuations	●●○	Sign forward contracts for organic coffee at start of season to set minimum prices
Market	Buying contract default	●○○	If international buyer fails to buy organic coffee, sell it as conventional coffee through local commercial houses
Default	Farmer sells coffee elsewhere	●●○	Contract obliges farmer to sell coffee to UCPCO UCPCO and FDL supervise harvest to assure delivery UCPCO offers best local market price and provides good client service UCPCO pays reasonable price for low-quality coffee
	No repayment	●●●	Check applicant's credit history and deny loans to defaulters UCPCO discourages defaulting by members
Currency	Exchange rates	●○○	Contracts include clause to maintain value against US dollar based on annual devaluation of córdoba of 5% UCPCO pays farmers for exported coffee in US dollars
	Inflation	●●● (long-term lloans) ●●○ (short term loans)	Value maintenance clause allows for 5% inflation

Risk levels: ●●● High ●●○ Medium ●○○ Low

negotiating position and can demand better terms from the buyers. FDL loans to UCPCO have risen from $40,000 in 2005 to $260,000 in 2008.

FDL The microfinance institution has used the opportunity to pursue with its mission by supporting a significant group of small-scale coffee producers. It has reduced its portfolio risk in several ways. Through the Union, it has better information about the farmers' performance. The Union's technicians visit the farmers regularly and advise them on how to solve production problems. The yields are more stable and quality is higher. Repayments are now handled by the Union rather than by individual farmers, and the Union will repay loans in full on behalf of the farmers. The risk is shared with the Union. FDL's collaboration with the Union also reduces its transaction costs. It can continue growing in the coffee sector and has a model it can use to develop alliances with other agricultural and livestock producers.

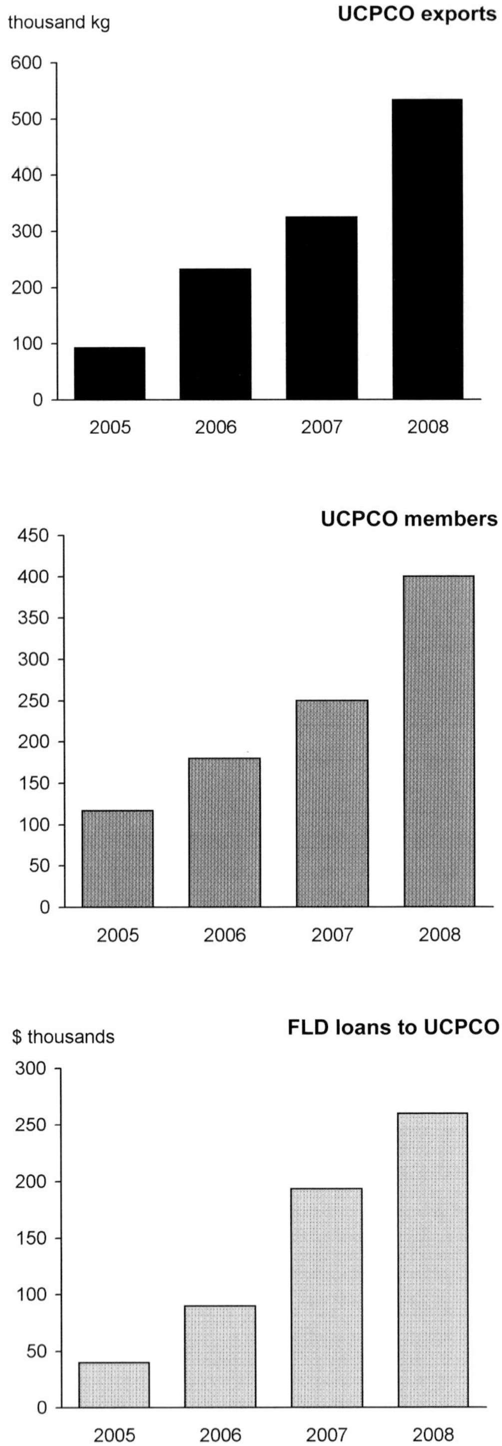

Figure 6.17 Increase in number of UCPCO members (top), coffee exports (middle) and FDL loans to UCPCO (bottom)

Lessons

Specialization The value chain has benefited because the Union, FDL and the farmers each focus on their own specialization.

- The Union has relinquished its role of providing working capital, so is able to focus on providing the services that it is good at, and that its members demand.
- FDL is able to provide microfinance services, and its alliance with the Union enables it to do so without incurring the costs of organizing and training the farmers.

- The farmers are able to focus on improving their yields and quality, without having to worry about whether they will get their loans on time.

Alliances Such specialization works because of the strong triangular alliance between the Union, the farmers and the microfinance institution. This alliance is advantageous to each of the parties. Their roles are complementary, and they share information transparently. The managers of the Union and FDL have been working together for a long time, so have developed a high degree of mutual trust and strong communication links.

Farmer empowerment FDL's previous attempts to finance coffee production failed because they did not take the producers' needs sufficiently into account. They were designed by FDL in collaboration with a trading house or a buyer, without involving the farmers. The farmers were at a disadvantage because they were obliged to deliver their coffee to the buyer, so were captive suppliers and could be easily exploited. The arrangement with the UCPCO is different: the Union negotiated on behalf of its farmer members, so their interests are reflected in the deal, and as co-owners of the Union, the farmers benefit from any profits it makes in the form of improved services, special projects and support such as scholarships, school supplies for children, and medical assistance.

Leadership The leadership of the Union and FDL managers has been key to success. Without their positive attitude towards change, commitment and ethical values, the chain would not have been as successful. It is, however, necessary to ensure that the sustainability of the chain does not depend on a few individuals.

Need for a kick-start The value chain was initiated through assistance from IDR, a government development agency (in other cases, the kick-start may come from an NGO, a development project, or the private sector). The initial grant and loan provided the capital, infrastructure and skills to develop the chain. Without such inputs, the chain would have developed more slowly, if at all.

Need for strong partners One of FDL's previous attempts to build a value chain involved a non-local buyer who did not have sufficient capital to sustain it. This chain ended in tears: some farmers are still waiting to be fully paid for the coffee they delivered. FDL is strong enough to bear the costs and risks, and has a local office in San Juan with other clients, so knows the area well and is not going to run away if problems arise.

The future

The Union wishes to get into the gourmet coffee sector. This is the highest value coffee. It can be auctioned among potential buyers, and may fetch nearly 90% more than conventional coffee (organic coffee, which the Union currently produces, fetches around 15% more). The price for gourmet coffee also tends to be more stable.

This will pose new challenges to the Union, and to FDL. The Union has identified certain farms above 1,200 m altitude that will be able to produce the quality of coffee required. Developing these farms will cost money, and the Union is asking FDL to provide new types of investment loans, with different disbursement procedures. The Union must also learn how to operate in the new, gourmet market.

For FDL, a challenge will be to maintain enough liquidity in the current international financial crisis. The flow of funds to microfinance institutions that rely on development finance organizations has been declining.

Given the success of the FDL–Union collaboration, FDL wishes to explore similar relationships with other cooperatives growing organic coffee. FDL now has a model to follow, but will have to invest in developing these new relationships.

More information

Alexandra Schutze*, Fondo de Desarrollo Local, alexschutze@gmail.com*

Manuel Bermúdez*, Fondo de Desarrollo Local, mbermudez@fdl.org.ni*

www.fdl.org.ni

Developing the organic quinoa chain in Bolivia

"Quinoa is one of my favorite meal staples - it's got more bite than rice, more protein than pasta. It's good hot or cold - and it's very easy to prepare."

So says an enthusiastic American blogger, one of a growing group of health-conscious Western consumers who have recently discovered quinoa, an ancient grain native to Bolivia and Peru. Health food stores are full of quinoa products, from organic whole grains to breakfast cereals, chocolate-coated quinoa bars and frozen quinoa dinners. Vegetarians buy it for its protein content; people with gluten intolerance love that it's gluten-free.

Figure 6.18 Quinoa is a staple food in the Andes

Quinoa produces edible seeds that are similar to cereal grains. Bolivians eat it on a daily basis. It is particularly suited to the growing conditions of the highlands in Peru and Bolivia because it can grow at altitudes of up to 4,000 metres. Over 10,000 farmers produce in total about 22,000 metric tons of quinoa per year on 32,000 ha of land. According to the National Institute of Statistics, the total value of quinoa exports reached $13.1 million in 2007.

Quinoa production is fundamental to the economy of rural communities of Bolivia's highlands, where it makes up 55 to 85% of farmers' revenues. The recent growth and dynamism of the quinoa value chain is largely the result of new export niche markets for foods with particular qualities: organic, nutritious, or originating from a specific region. These have emerged in response to changes in demand, as Western consumers have become more concerned about their health and the environment.

Actors in the value chain

Irupana Andean Organic Food (Irupana) is a private company that has been processing and exporting Andean cereals (quinoa and amaranth) to international markets since 2003. Irupana is based in the Bolivian capital, La Paz, and sources its quinoa from farmers and collectors in five municipalities (Salinas de Garcí Mendoza, Pampa Aullagas, Quillacas, Colcha K and Uyuni) in the southern Altiplano (Figure 6.20). This chapter presents the experience of Irupana in working with the farmers of these municipalities in developing a business relationship in 2003–7.

Figure 6.19 shows how the quinoa value chain was organized in 2003. There were three major actors, farmers, traders or "collectors", and Irupana.

Small farmers In 2003, quinoa was grown by subsistence farmers and their families. On average, they had a production area of 0.25 hectare, yielding about 400 kg annually. They consumed most of this themselves, leaving a small surplus to sell. The farmers sold directly to local collectors, or to collectors who worked

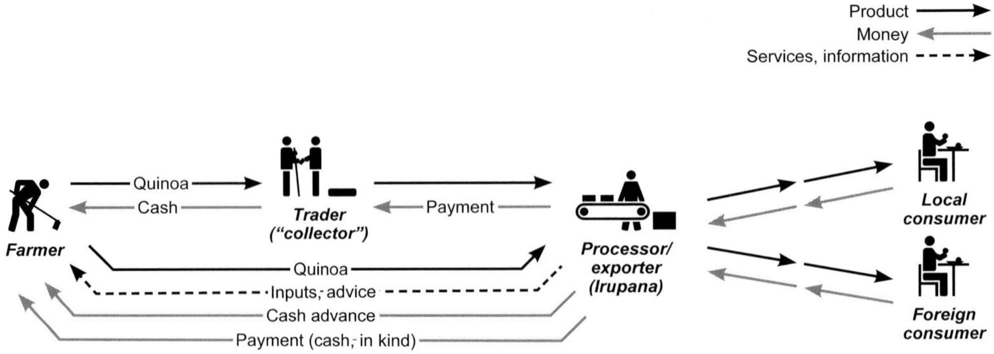

Figure 6.19 The quinoa value chain in the southern Altiplano, 2003

for farmers' associations or private enterprises that buy large volumes of quinoa year-round. The farmers received payments in cash or in kind (seeds, food). Low soil fertility limits productivity. With almost no roads in the area, and without a secure market outlet, there was little incentive for the farmers to try to increase production. They also had no technical assistance and limited access to market information (prices, volumes and other related issues). Their total annual income was about US $1,684 from various sources (selling potatoes, quinoa or working occasionally as labourers).

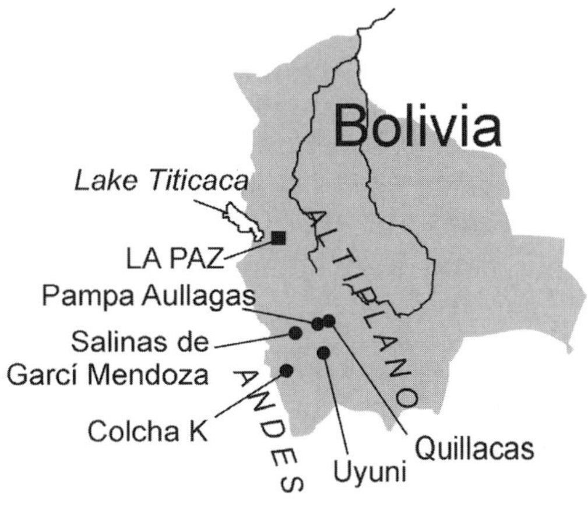

Figure 6.20 The quinoa production area that supplies Irupana

The stages in the quinoa harvest cycle include sowing, harvesting, threshing, winnowing and drying. Sowing begins in September and can last until the end of October. Abundant rains in January, February and March allow the crop to grow fast and yield well. Harvesting starts in March/April and is done by the whole family. Winnowing is done by hand using a type of sieve and shaking it in the wind.

Traders Traders, also known as "collectors" or *acopiadores,* go from door to door to buy dried quinoa from farmers they have verbal agreements with. These agreements cover the quantity, quality, forms of payment, price and delivery terms. The traders operate in areas where farmers do not have their own transport. Farmers with a lot of grain to sell have almost no choice but to sell it at the prices fixed by the traders. Some traders also buy from local markets. They are well-informed about market conditions and they have good contacts and relationships with farmers, farmers' associations and production companies. Their working capital ranges from $5,000 to $50,000, and consists of their own savings or money borrowed from family members.

Irupana Irupana was started in 1987 by an entrepreneur with a strong sense that organic products were the way of the future. The company began by stimulating organic coffee production in the Irupana region (about 70 km east of La Paz), and grew to the point that it offered about 80 different products. Eventually, the company ran into financial difficulties, partly because its product range was so wide. In 2003, the company underwent a major reorganization. It cut back on its product range and decided to focus mainly on organic quinoa and amaranth for export.

Why did Irupana decide to focus on organic quinoa? The company carried out 5 years of research into the demand for organic quinoa in Europe and the USA, and hired two international experts to prepare a marketing report. It participated in several major international trade fairs in Germany, France and the USA to explore the market. It found that international demand was almost exclusively for certified organic quinoa. The prices for organic quinoa were rising every year in response to the growing demand for health products. In 2005, the organic quinoa market (Bolivian and export market combined) was worth $5 million. Prices for organic and fair trade quinoa were about $1,200–1,300 per ton. It is hard to estimate how stable this market is, because so far demand has continued to increase. New markets are opening up, such as in Venezuela. By June 2008, the price was $107 per quintal (1 quintal = 46 kg), or above $2,300 per ton.

Irupana now produces organic quinoa products such as cereals and bars for the local market, and bulk organic quinoa for the export market. The export of Andean cereals has become its core business, and the company's growth depends on it being able to source the raw materials. In 2003, increasing its production of organic quinoa products meant finding more farmers, in addition to its regular suppliers, who would be willing to become certified for organic production.

Irupana buys its raw material from traders, but it also has verbal agreements to buy from small farmers directly. The agreements with farmers cover the quantity, quality, forms of payment, price determination and delivery methods. They sometimes include the provision of some production inputs (such as seeds and organic insecticides) to farmers, a cash advance to help finance production, and technical advice on organic production.

Figure 6.21 Quinoa is used in a wide variety of health foods

Going organic

For a company to be able to meet consumer demand for organic products, it needs to be able to rely on regular suppliers. Because organic standards are so demanding, there tends to be a limited number of producers who can meet these standards. Getting organic certification is a complicated and expensive process. Buyers and consumers of organic products demand a lot of information to assure that the product is truly organic. This means that the chain needs to be highly integrated, with good information flows all the way down the chain.

Irupana's decision to start trading in organic quinoa products only meant greater demand for organic quinoa, and therefore changes in production processes and in the value chain. In order to expand organic production, a number of problems needed to be addressed:

Farmers

- **Costs of certification** The certification process, in which specialists check that farmers use the right inputs and treat the land according to organic standards, can take 3–5 years. The costs of this process can be prohibitive for small farmers with little working capital. For example, individual certification costs about $300 per hectare, which is almost 20% of the farmer's yearly income.
- **Costs of organic production** The costs of changing to an organic farming system are high for small farmers. Organic farming requires special inputs, such as biological control agents and bio-insecticides, which require special equipment to apply. Farmers also needed to invest in upgrading their post-harvest facilities, for example by getting new grading and winnowing equipment.
- **Knowledge and information** The organic market's strict requirements related to quality, certification, and transparency could not be easily met by small farmers. The farmers had little knowledge about appropriate technologies and no access to technical assistance. They wanted a firm contract and secure prices.

Traders

- **Access to working and investment capital** The traders that Irupana works with needed capital to invest in making their storage facilities compliant with organic standards. They also needed working capital to pay farmers in cash for their organic quinoa.

Irupana

- **Matching supply with market demand** Irupana faced a growing demand for its organic products from the export market. Small farmers, cultivating an average production area of 0.25 ha, had average yields of just 400 kg a year. Irupana calculated that it needed about 2,000 kg per farmer to satisfy market demand, and would need to increase the number of farmers supplying certified grain.

- **Lack of working capital** Irupana's decision to focus on organic products was sound – organic production is a profitable business – but it required major investments. Following the company's reorientation towards organic quinoa and amaranth, it spent a lot of money on improving the existing plant and buying the equipment it needed. As a result, Irupana no longer had the working capital to expand their operations and meet the growing market demand.

Irupana approached several banks looking for a loan, but its financial statements at that time were not attractive to commercial banks. Moreover, the new business model had not yet proven itself. Irupana needed another type of financial institution to come on board: one that does not look at the past (the financial statements), but rather at the future (the business plan). Irupana needed a financial institution that believed in the business and that would be willing to become a risk-taking partner.

Bringing innovation to the chain

In 2003, Irupana decided to approach Pro-rural, a financial institute that supports small and medium enterprises in Bolivia (Box 6.5). It explained that it had an unmet market demand for organic quinoa and wanted to start sourcing organic quinoa from small farmers in the south Altiplano. Pro-rural agreed to help strengthen the value chain under its "temporary joint venture" fund. Starting in 2004, Pro-rural began providing support to the three main actors in the quinoa chain.

Temporary joint venture with Irupana

Temporary joint ventures (Box 6.6) are suitable for financing start-ups that lack working capital, hard collateral and experience in business. As a company that had just changed its core business, Irupana was in a similar position.

The first such joint venture between Pro-rural and Irupana was established in 2004 for a period of 14 months, with a total investment of $150,000 from Pro-rural (60%) and slightly over $100,000 (40%) from Irupana. The funds were to be used to buy organic quinoa from suppliers and sell it on the export market. The 14-month time period was related with the commercial cycle of the business. The operation was successful, and Pro-rural received a return on investment of 8.7%.

A second, 36-month, temporary joint venture was set up in 2005. This time Pro-rural invested $250,000, giving it a 48% share in the company, and Irupana contributed $272,000 (52%). As minority shareholder, Pro-rural reduced its risk. The return on investment was 8.1%. Both these financing operations allowed the company to capitalize and improve its financial performance and statements.

Box 6.5 Pro-rural

Pro-rural is a financial institution that specializes in working with small start-up businesses that are generating revenues but are not yet profitable. It provides them with necessary capital and organizes technical assistance, such as support in managing the business or setting up an accounting system.

Pro-rural believes that intervening with financial and non-financial services will reinvigorate rural economies. It believes that small and medium enterprises and small farmers' associations have a critical role to play in the Bolivian economy. It brings the principle of risk capital (usually associated with high interest rates and big business) to address rural development issues, providing seed money, start-up capital or starting temporary joint ventures.

Pro-rural's temporary joint ventures fund (a type of venture-capital fund) was set up 2001 to provide enterprises with early-stage capital. This fund has an investment portfolio of $1.5 million, a $4.42 million credit portfolio, and seed and start-up capital of $600,000.

More information: www.prorural.org.bo

These partnerships (❶ in Figure 6.22 and Table 6.7) enabled Pro-rural to provide financial inputs without giving Irupana a loan, so avoided burdening the company with a large debt in its financial statements, and helping with cash flow. The risks were also shared. Pro-rural had a vested interest in making sure the business would succeed. As a shareholder in the joint venture, Pro-rural had insights into what was going on in the business and a major say on the transactions it was partially financing.

Temporary joint ventures with traders

Pro-rural used a similar approach by going into temporary joint ventures with traders (❷ in Figure 6.22 and Table 6.7). One example is Don Lino, a small-scale producer who decided to become a trader as well (Box 6.7). Pro-rural formed a temporary joint venture with him for a period of 10 months: Don Lino contributed $20,775, while Pro-rural invested $20,000, taking a 49% share in the enterprise. Don Lino used the money to buy organic quinoa, which he then sold to Irupana for export. After 10 months, he and Pro-rural shared the profits; the return on investment was 9.8%. The initial investment was converted into a loan, which Don Lino repaid over a 3-month period at an interest rate of 8%. Pro-rural repeated this approach with three other traders as well.

Start-up capital and technical assistance for farmers

The Inter-American Foundation (a US government donor agency) donated $300,000 for Pro-rural to invest in infrastructure, equipment, technical assistance and organic certification for farmers (❸ in Figure 6.22). Of this sum, 41% was spent on fixed assets such as winnowing and grading equipment, weighing scales and silos. This was a 3-year programme from 2004 to 2007; in the first year, a farmer could get a grant covering half the cost of a new machine. In the second

Box 6.6 Temporary joint ventures

Venture capital is equity investment made, typically in less mature companies, for the launch, early development or expansion of a business. Conventional venture capital focuses on businesses that need over $1 million and that target urban areas, services or exports. Smaller scale businesses such as Bolivia's rural and semi-urban entrepreneurs cannot find the risk capital they need. Traditional credit is too restrictive: it requires hard collateral and entails monthly repayment obligations and high interest rates.

Because most of Pro-rural's clients are small-scale farmers and their associations, Pro-rural started looking for an alternative to traditional loans. In 2001, It started testing temporary joint ventures. This approach has been used by three other institutions in Bolivia, but Pro-rural is the only one to engage in temporary joint ventures with small-scale farmers and their organizations.

The term "temporary" refers to the fact that the joint venture is set up for a short, fixed period. The period depends on the commercial cycle of the business: typically, it may last one business cycle, from when the trader first buys the crop to when he or she has sold it to the next actor in the chain. Through a joint venture agreement, Pro- rural and its partner combine their expertise and resources to develop a particular business. It may be thought of as a partnership that exists only for this specific project. The parties involved maintain their separate business and financial identities, and they do not register as an enterprise.

The parties need clearly defined terms in order to avoid unpleasant surprises in mid-project. The agreement must provide a comprehensive road map of the duties and obligations of both the parties and also the profit shares. The business plan is the key instrument to follow up the operation.

A joint venture involves sharing money and risks, and sometimes sensitive information. Pro-rural's involvement is not just financial. Whenever necessary, Pro-rural adds value to the partnership by bringing in technology, accounting software or technical assistance projects. All these factors contribute to the strength and sustainability of the business, so decrease risk. Pro-rural also supports community development: for example, it helped Irupana import 300 solar panels from India for highland communities.

One farmer sums up the Pro-rural's contribution: "Why do I have to share the profit with Pro-rural, which just sits there? Then I realized that they were my partners, they were investing their money, time, technology and know-how...They were just like me, or maybe more than me."

year, the same farmer could qualify for a second grant covering 30% of the cost of more equipment.

Some 36% of the $300,000 was spent on co-financing technical assistance to farmers to overcome major bottlenecks in productivity, quality management and certification. To get this assistance, the farmers were required to make a 50% contribution to the cost. The assistance was provided by Irupana technicians, who were able to advise farmers on which equipment to acquire because they were familiar with the situation on the ground.

Irupana also offered the farmers a 20% premium for organic quinoa to ensure they met the quality standards of organic production and to retain them as regular suppliers in a very competitive market.

Dissolving the partnership

In February 2009, Pro-rural stepped out of its partnership with Irupana. Irupana had become bankable; it is now a solid company. It anyway now needs more capital than Pro-rural can provide. Irupana had some interest in a permanent joint venture, but Pro-rural is not interested at this time because long-term investments require a different type of know-how.

Risk analysis

Why did Pro-rural trust Irupana to manage the joint venture well, given that the firm was in financial difficulties when it came to Pro-rural? Why did Pro-rural not consider it necessary to give Irupana technical assistance, in financial management for example?

Irupana was able to provide Pro-rural with soft collateral in the form of export and purchase contracts. In the case of the joint venture with the collector Don Lino, Pro-rural had a purchase contract from Irupana to Don Lino as soft collateral. In addition, Pro-rural knew that selling organic quinoa was a good business to be in. Pro-rural first looks to see if a business shows promise. In the case of the farmers, technical assistance was necessary to increase their productivity and quality.

Pro-rural's financial interventions are not secured by hard collateral. Instead, it looks at the performance of the value chain as a whole, weighing the strengths

Box 6.7 Organic quinoa revives the local economy

"Before Irupana came here, the town was dead. Now we have revived."

Doña Lidia, an Aymara woman from Sevaruyo, grows organic quinoa with her husband, Don Lino, in the southern Altiplano. "Growing quinoa was not secure. We were never sure how much we would earn, so we just grew enough for our family. A lot of people left the area, but now they are returning here to work and benefit from the good prices for organic quinoa."

For these small-scale farmers, growing quinoa without chemicals was natural. "It was what our parents did", says Doña Lidia. "With conventional quinoa production, the *Pachamama* (mother Earth) suffers within 3 years. We fumigate with organic products to defend our *Pachamama*."

Her husband, Don Lino, decided to go into business as a collector, in addition to growing quinoa. He approached Pro-rural with his idea. Pro-rural set up a temporary joint venture with Don Lino so that he would have access to working capital to buy quinoa to trade.

Don Lino and Doña Lidia also benefited from the Pro-rural/Inter-American Foundation start-up capital fund: they bought a grading machine.

"The grading machine was expensive. We were able to buy it because we contributed half of the cost price. Now, we also have a quinoa depot", explains Doña Lidia. "Production is increasing, we have secure markets, and many of us have already bought tractors."

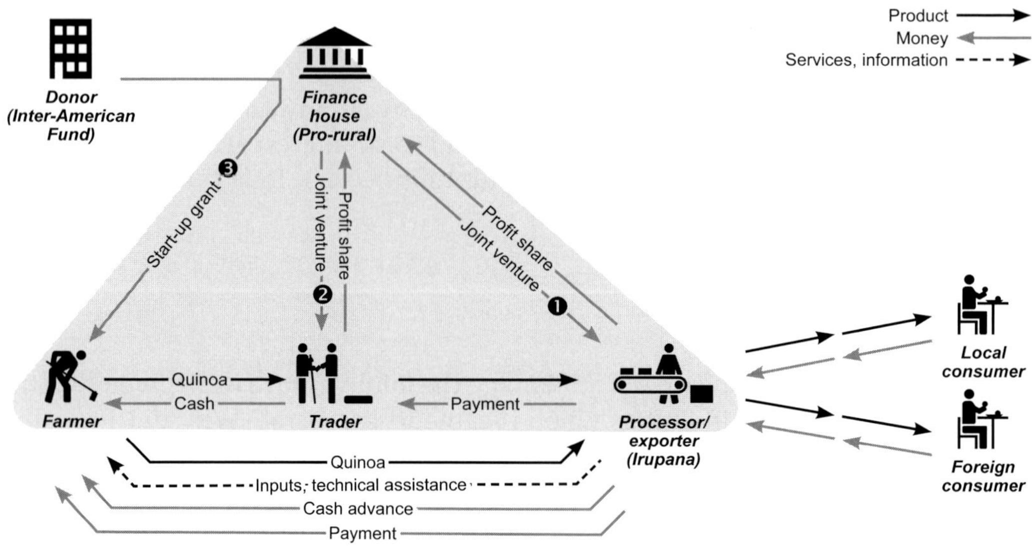

Figure 6.22 The quinoa value chain after Pro-rural's intervention

of the chain against the risk. The key factors that help Pro-rural control the risk in the organic quinoa chain are:

- As a partner in the joint ventures, Pro-rural became a chain actor. It shared in the risk and had access to all the information necessary to monitor the performance of the chain. Pro-rural had a say in decision-making and an inside view of what was really going on.
- Pro-rural was involved at all three levels of the chain: at producer, trader and processor level. It had information on performance and progress at all stages.
- Irupana paid an additional price premium of 20% above the spot market price, earning the loyalty and commitment of its suppliers.
- Irupana's management demonstrated its capacity to deal with the problems and risks of quinoa production and export. Irupana maintained close relationships with the other actors in the chain.

Table 6.8 summarizes the risks involved in the quinoa value chain.

Benefits

Farmers The average farmer's annual income has risen dramatically. Production and quality have both increased. With an assured market, each family now produces an average of 8 ha of quinoa (compared to just 0.25 ha before). Yields average 0.78 t/ha (though may vary from 0.4 to 1 t/ha). The average price is Bs 12 per kilogram, resulting in a gross income of Bs 74,800 a year in 2007, or $10,389, compared to just $1,684 from all sources in 2004.

Table 6.7 Pro-rural's joint ventures with Irupana and a quinoa trader in Bolivia

	Pro-rural–Irupana joint ventures (❶ in Figure 6.22)	Pro-rural joint venture with trader Don Lino (❷ in Figure 6.22)
Product and financial flows		
Purpose	First joint venture (2004): Buy organic quinoa and sell it for export Second joint venture (2005–8): Store and sell organic quinoa	Store and sell organic quinoa
Amount	First joint venture: $250,000 (60% Pro-rural, 40% Irupana) Second joint venture: $522,000 (48% Pro-rural, 52% Irupana)	$20,000 (49% Pro-rural, 51% trader)
Period	First joint venture: 14 months Second joint venture: 36 months	10 months
Disbursement	Each party maintains own account	Each party maintains own account
Repayment	Profits shared at end of joint venture period	Profits shared at end of joint venture period
Interest rate	Return on investment: First joint venture: 8.7% Second joint venture: 8.1%	Return on investment = 9.8%
Risk management		
Securitization	Export contracts	Contracts with Irupana
Liability	Joint	Joint
Information flows		
Information required to apply	Sale contracts (business with growth potential) Previous year's sales (at least $20,000) Minimum contribution by partner of 51% Net assets similar or bigger Business plan, identity documents Authorization by partner's Board, legal representation of the firm specifically for joint venture Legal bureau report detailing partner's debts and overdue payments	Sale contracts (business with growth potential) Previous year's sales (at least $20,000) Minimum contribution by partner of 51% Net assets similar or bigger Business plan, identity documents Authorization by trader, legal representation of the firm specifically for joint venture Legal bureau report detailing partner's debts and overdue payments
Information required during season	Monitoring reports (business plan) Investment agreement	Monitoring reports (business plan) Investment agreement
Time lag between application and payment	1–2 months	1–2 months

Traders In the same period, the trader Don Lino Valez increased his volume from 275 tons to 483 tons, and sales from $90,000 to $150,000.

Irupana Irupana increased the amount of quinoa it handled from just under 700 tons in 2003 to nearly 2,600 tons in 2005. Its sales rose from $210,000 to $810,000 in the same period, and its assets rose by 70%, from $890,000 to $1.5 million. Irupana became bankable: it can now get loans from commercial banks, and has risen from being the fifth largest exporter of organic quinoa in Bolivia to second place.

Pro-rural. Pro-rural had a good client in Irupana and the traders. Irupana has graduated from its need to rely on Pro-rural, and the joint ventures have been completed. Pro-rural continues to have smooth working relationships with quinoa traders.

The chain The number of farmers involved in organic quinoa production increased from 217 in 2004 to 391 in 2007. The total production of organic quinoa tripled from 10,500 quintals in 2004–5 to 29,000 quintals in 2006–7.

Challenges

Pro-rural is no longer involved in this value chain. The challenge for the chain actors is to continue to function at the same level of efficiency as during the 4 years of Pro-rural's intervention.

Pro-rural is planning to replicate the model in other value chains: for wild cocoa, amaranth, sesame and wool.

Pro-rural is also replicating the model of value chain financing with farmers' associations and indigenous groups so it can compare the experiences and identify the key elements to increase the value share of indigenous farmers. A challenge will be to maintain innovativeness and efficiency. The idea is that farmers' associations would look after product processing, marketing and exports. Four farmers' associations have approached Pro-rural with a request to go into a joint venture to set up a rural investment fund. One of these is a quinoa farmers' association.

Lessons

Suitability of temporary joint ventures Temporary joint venture capital is suitable for financing a start-up company where there is a lack of working capital, hard collateral and experience with business. The business has to have clear potential for growth, and the partner has to be able to put up at least half of the capital required. Subsidies in the form of grants from the Inter-American Foundation also played a role in the development of the organic quinoa chain and may be necessary for some start-up initiatives.

Table 6.8 Risk analysis for the quinoa chain in Bolivia

	Source of risk	Risk level	Mitigation strategy
Production	Crop quality, quantity and productivity	●●●	Project financed by Inter-American Foundation
Price	Market price fluctuations	●●●	Price fixed when contracts are signed
Default	Farmers do not honour agreement	●●○	Close relationship with partners
	Sales to third parties	●●●	20% premium paid to farmers by Irupana
Currency	Exchange rates	●○○	
Other	Farmers demand more benefits or better conditions	●●●	Close relationships with farmers' leaders Social investment in communities (such as supplying solar panels)

Risk levels: ●●● High ●●○ Medium ●○○ Low

Financial and technical support mix The combination of financial and non-financial services such as technical support met the needs of the actors in the organic quinoa value chain.

Becoming a chain actor By taking on the role of co-owner, an investor increases the chances that the chain will be successful. A co-owner participates in decision-making and assumes risks and challenges, in addition to investing resources.

Flexibility Both Pro-rural and Irupana had to adapt to changing conditions and learn how to work as part of the joint venture.

Duration It took Pro-rural about 3 years of partnership with the quinoa chain actors to achieve results. This period was necessary to see change and the impact of the intervention, and to allow chain relationships to stabilize.

More information

***Maria Renée Bejarano**, renee.bejarano@gmail.com*

PRO-RURAL, prorural@prorural.org.bo

www.prorural.org.bo

Financing artisanal fishing in India

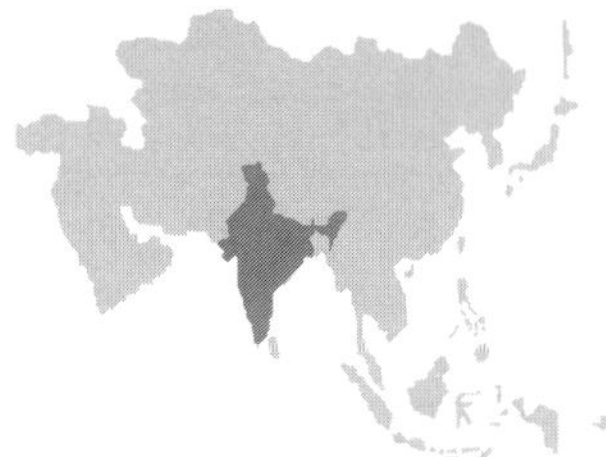

DRAGGING THE BOAT ONTO the beach at Tharangambadi is hard work. It was a good catch last night, and the boat is full of fish: big, silver, seer fish, weighing perhaps 5 kg each, as well as smaller skipjack tuna, around 2 kg each. There is nearly a ton of fish in the bottom of the boat. A crowd of villagers gathers in the early morning sun to help the five crew pull the boat up the sand and unload the fish into crates. Soon there is a stream of people carrying crates of fish up the beach to a building on the shore. They heave the crates onto an electronic scale, and a man notes on a form the type of fish, the weight and the owner.

Tharangambadi is one of many fishing villages that dot the coast in Nagapattinam district, in Tamil Nadu, southern India. The building belongs to the village fishermen's society, one of such 150 societies affiliated to the Trivandrum-based South Indian Federation of Fishermen Societies (SIFFS, Box 6.8). When the fish

Figure 6.23 Smaller fish are auctioned on the beach (left). More valuable fish are taken for weighing before they are sold (right)

Box 6.8 South Indian Federation of Fishermen Societies (SIFFS)

SIFFS is the apex body of a three-tier structure of autonomous organizations of small-scale fisherfolk. These organizations are based in coastal districts in Karnataka, Kerala, Tamil Nadu, Pondicherry and Andhra Pradesh in southern India. SIFFS is the result of major church NGO initiatives that began in 1970. A large number of NGOs worked independently and in concert to create new cooperative institutions for the fishermen. SIFFS developed in several phases:

1970–80 The first phase saw the creation of a model village society that gave the fishermen control over beach-level fish marketing and enhanced their incomes through better fish prices and reduced outflows to moneylenders and middlemen. This model gradually spread to neighbouring districts. At this time, SIFFS facilitated its members to get small loans through microfinance institutions.

1980–90 The district and regional federations were formed, followed by the apex organization, covering the whole of southern India, in 1982. These higher-level structures enabled greater economies of scale in managing inputs and markets. They also developed and promoted new technologies such as improved boats and motors. SIFFS also developed revolving funds and added boat financing to its services.

Since 1990 Boatyards, outboard motor services and ice plants have been developed, and SIFFS has started to provide credit. In 1996 SIFFS took up "livelihood financing" to free the fishing community from their reliance on moneylenders, middlemen and merchants. It then started offering other financial services such as investment and working capital.

SIFFS's services reach nearly 100,000 families, or a total population of about 400,000 (Figure 6.24). At the base level are over 150 **village fishermen's societies** that function like cooperatives. They market fish caught by their members, and cater to the financial requirements of over 45,000 fisherfolk. SIFFS serves 9,000 fishing boats, each with 4–5 crew.

SIFFS has 10,000 women members, many of them fish vendors, grouped in over 170 **fisherwomen's societies.** It offers them credit and post-harvest technical support.

The fishermen's societies are affiliated to five independent **district or regional federations** which monitor and support the village activities. They also undertake a number of commercial, technical, educational and welfare activities. The fisherwomen's societies are affiliated to women's federations which work in partnership with the men's federations.

At the **apex level**, SIFFS focuses on technology for small-scale fish workers and assists in the coordination and management of the district federations. One example is a marine plywood boat produced and disseminated by SIFFS after a decade of research and development. This design is stable, even in high waves, can land on the beach, and can be equipped with an outboard motor (another major improvement to the fishing techniques).

The apex body collects market data and publishes information on the small-scale fishing sector. It also focuses on the marine fish resource management and safety for small-scale fishermen. The government covers the costs of a pilot market information service. Donor grants have funded research and development activities and policy research, but SIFFS is increasingly paying for these activities out of its own funds.

SIFFS collaborates with government agencies such as the Marine Products Export Development Authority, the Central Marine Fisheries Research Institute, and the Central Institute of Fishing Technology. These provide services such as information on improved technologies and fishing trends, export promotion, and subsidized crates and ice boxes.

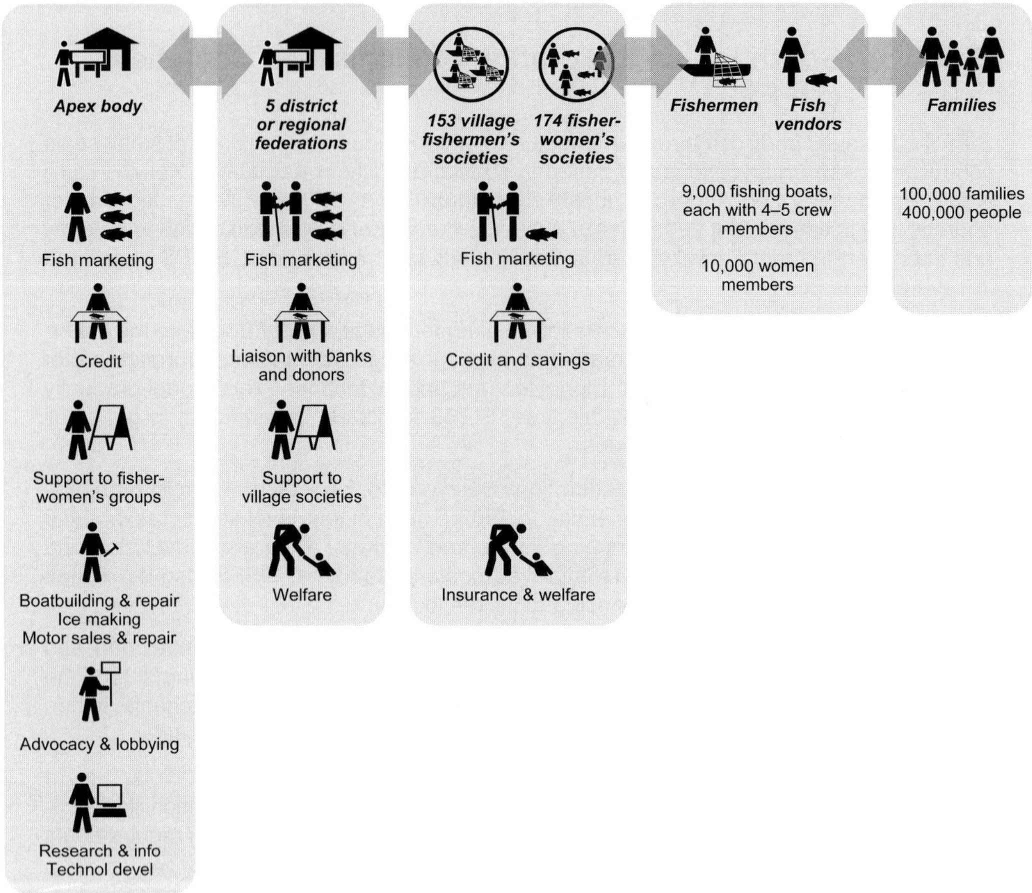

Figure 6.24 Services provided by different levels of SIFFS

have been weighed, society employees grade and sort the fish by species and size. They spread a layer of crushed ice on the concrete floor and lay the glistening fish down on top. They add another layer of ice, then more fish, and so on until there is a big mound of fish in the middle of the room.

Auctioning fish

The merchants have started arriving to inspect the fish. There are ten merchants in the village, well known to everyone present, and they all want to buy fish to take to Chennai, 400 km along the coast to the north, or even to Bangalore or Trivandrum, both 700 km away. They already know how much fish of what type is available: the society staff have informed them.

At 8 o'clock, the society auctioneer takes charge. He starts off with the bigger seer fish, above 3 kg. "Who will offer 150 rupees per kilo?" he calls. The merchants start bidding against each other. Someone offers Rs 160, then another merchant bids Rs 180. The bidding goes on until the final price of Rs 270 ($5.40) – a good price. Then the auctioneer sells the smaller seer fish, then the tuna.

The merchant's workers load the fish and more ice into a lorry, and by 10 o'clock, the lorry is ready to leave for the city wholesale market.

When all the fish have been weighed, the owner takes the completed form and his society account passbook into an office in the same building. The clerk there calculates how much the fish are worth, deducts 10–15% to cover loan repayments and 5% to cover expenses. This 5% consists of 2% to cover the village society's expenses, 1% for the district federation, and another 2% for compulsory savings that goes into the fisherman's account.

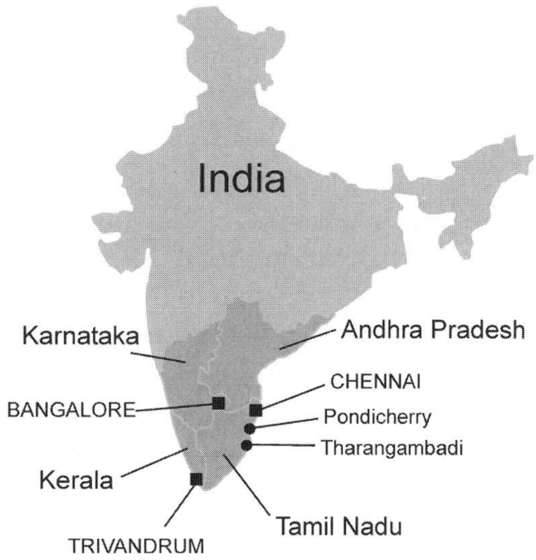

Figure 6.25 SIFFS serves fisherfolk in coastal areas in southern India

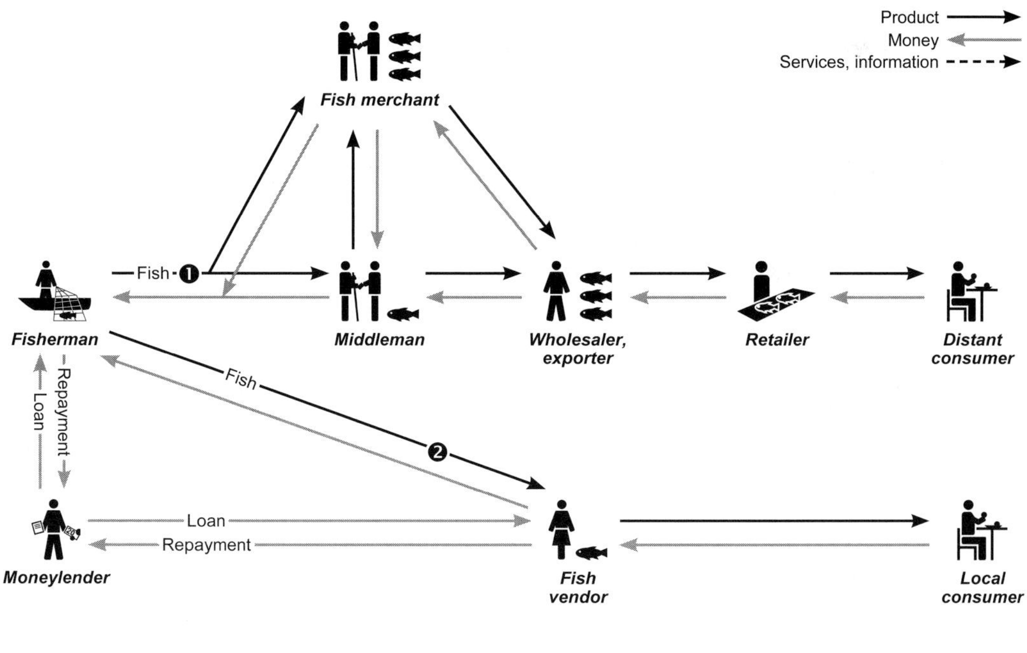

Figure 6.26 Fish value chain before SIFFS's intervention

The clerk makes a note of the transaction in the passbook and in the society's account book. Then she opens the strongbox by her desk and carefully counts out a wad of rupee notes and gives them to the owner.

On some days none of the merchants bids the society's minimum price. Then the society itself will market the fish. SIFFS staff collect information on prices in major wholesale markets every day, and the village society can call to find out today's price, or check it on the SIFFS website. If the price is right, the society will hire a lorry and take the fish to the wholesale market. The merchants know this, and they can check the prices on the website themselves, so it is fairly rare - perhaps twice a month - that the society has to market the fish.

Seer fish and tuna are table fish that fetch a premium price. For smaller or lower-quality fish such as mackerel, sardines and anchovies, there is an auction on the beach. Women vendors gather round the piles of fish and bid for lots (the fish have not been weighed, but everyone is good at judging how heavy they are). The society auctioneer manages the sale: he notes the names or each buyer and seller and the amount agreed, and collects the money from the buyer. The fisherman selling the fish gets a chit with this information, then goes to the office to pick up his cash (minus the same deductions as for the premium fish).

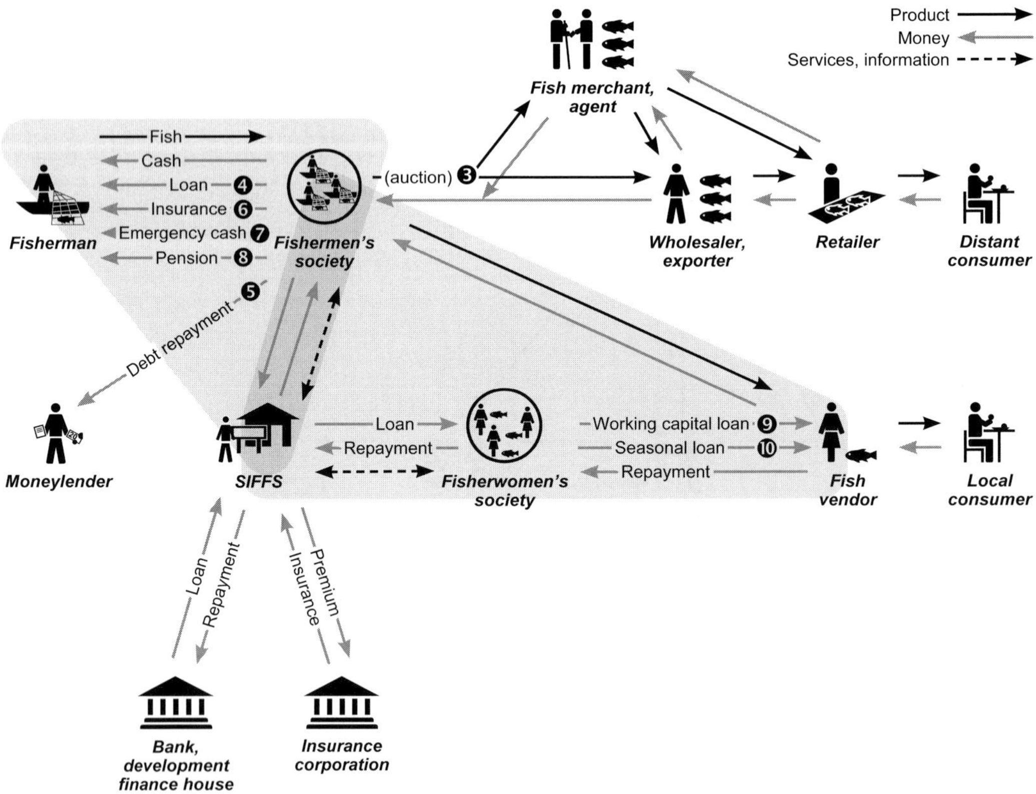

Figure 6.27 Value chain after SIFFS's intervention

The fish value chain

Fishing is very important in India. The fisheries sector is the source of livelihood for 900,000 fishermen and 800,000 workers in allied activities in poor fishing villages, and provides jobs and income to many others not directly involved in fishing. It produces cheap, nutritious food for millions, stimulates related industries, and generates foreign exchange.

Marketing fish in Tharangambadi and other villages was not always like this. In the old system, before the society got involved in fish marketing, the value chain had seven main stakeholders: fishermen, middlemen, merchants, wholesalers/ exporters, retailers, fish vendors and consumers. The fishermen sold their catch individually to middlemen (local people who advanced money to the fishermen in return for taking control of the catch when it was landed) or merchants (bulk traders who bought fish on the beach). The middlemen and merchants sold on the fish to wholesalers or exporters, who in turn sold to retailers, who marketed to the consumers. This chain is shown as ❶ in Figure 6.26.

The fishermen also sold their catch to fish vendors (❷ in Figure 6.26). Most of these are women - often the wives and daughters of the fishermen. They sell these fish in the local markets, or carry them in baskets to surrounding villages to sell them door to door.

This system was underpinned by moneylenders: local people who gave loans to the fishermen and the fish vendors at interest rates varying from 36% to 120% a year, or in return for 10% of the catch just to pay the interest on the loan. They also lent to middlemen and merchants at lower interest rates.

This system severely disadvantaged the fishermen and women vendors. The interest rates and fees charged by moneylenders were high. Without a conventional source of investment credit to pay for boats or to maintain equipment, and with no working capital, many borrowers found themselves trapped in debt. Their income varied with the size of catches and the daily prices. Fishing is a dangerous occupation, with accidents and deaths all too common, and the fishermen had no form of insurance. SIFFS was established to help fishermen overcome these problems.

Revising the value chain

The new value chain built by SIFFS (Figure 6.27) empowers the fishermen and the women vendors through a series of financial services. Instead of the moneylenders, SIFFS provides various forms of finance to village societies of fishermen and -women. It is able to do this because it can access loans from development finance organizations and commercial banks. The whole system is supported by information flows managed by SIFFS.

SIFFS and its member societies offer seven types of business service: organized fish sales, debt redemption, production loans, women's working capital, seasonal

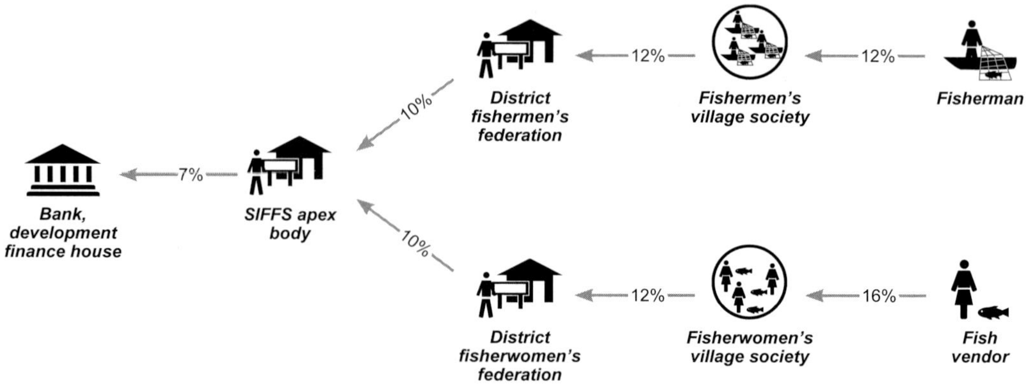

Figure 6.28 Interest rates paid for loans by different levels in SIFFS

loans, insurance, and an old age security savings scheme (Tables 6.9 and 6.10). We will examine these one by one.

Organized fish sales

The auction system described above gives the fishermen better prices than they would get by selling individually. The system is transparent (everyone sees how much the fish weigh, and anyone can bid to buy), and payment is immediate (❸ in Figure 6.27).

Under the new system, the middlemen may buy on behalf of people who cannot attend the auction. If they do this, they take 10–15% commission on the sale value.

Investment loans

Fishermen's assets tend to have a short life. Nets tear, boats leak, engines break down. The village society loans money to individual fishermen to cover these costs – between Rs 10,000 and Rs 100,000 ($200–$2,000). A fisherman is allowed to borrow up to 25% of the value of his previous year's catch and is required to repay the principal plus interest (12%) within 42 months (❹ in Figure 6.27 and the first column of Table 6.9).

The borrower must begin repaying the loan immediately: the society deducts 10% of the value of the daily catch. If the catch is lower than expected, the repayments are correspondingly lower. Each month, the society calculates whether the borrower has repaid enough. If not, he can increase the amount deducted (say, 15% of the catch instead of 10%) or can make up the balance at the end of the month.

Interest is calculated on the balance of the loan remaining. The repayment rate is assessed on the basis of the 36-month repayment period on an equal monthly instalment basis.

Table 6.9 SIFFS loans to members in India

	Fishermen's investment loan (❹ in Figure 6.27)	Fisherwomen's working capital loan (❾ in Figure 6.27)
Product and financial flows		
Purpose	Acquire/replace fishing equipment	Working capital for fish vending and petty trades
Amount	Rs 5,000–100,000 ($100–$2,000) 9,715 loans totalling Rs 269.3 million ($5.4 million) Average = Rs 28,000 ($560)	Rs 2,000–30,000 ($40–$600) 8,336 loans totalling Rs 50.4 million ($1 million) Average = Rs 6,000 ($120)
Period	42 months	12–24 months
Disbursement	Directly to borrower through village society	Directly to borrower through village society
Repayment	36 instalments	Daily
Interest rate	12%	16%
Transaction costs	Upfront fee: Rs 20 ($0.40) per loan	Upfront fee: Rs 10 ($0.20) per loan
Risk management		
Securitization	Borrower must be active member of society Formal collateral not required; informal social pressure used to assure repayment As last resort, boats or equipment may be confiscated	Borrower must be active member of society Formal collateral not required; informal social pressure used to assure repayment
Liability	Society and individual borrower	Society and individual borrower
Information flows		
Information required to apply	Application form Society's assessment of individual needs Assessment of previous 12 months' catch History of previous loans Screened at village, district and apex levels before funds are transferred from apex to village funds for disbursement to applicant	Application form Society's assessment of individual needs History of previous loans Screened at village, district and apex levels before funds are transferred from apex to village funds for disbursement to applicant
Information required during season	Monthly financial statement contains borrowers' catch details and financial status	
Time lag between application and payment	Under 2 weeks	Under 2 weeks

Table 6.10 Other selected SIFFS financial services in India

	Debt redemption loan ❺	Insurance (group basis) ❻	Seasonal loan ❿
Product and financial flows			
Purpose	Repayment of debt with middlemen	Death, permanent disability	Working capital for women with excellent performance histories
Amount	Rs 5,000–40,000 ($100–$800)	Rs 75,000 ($1,500): death or permanent disability Rs 35,000 ($700): partial disability 8,821 beneficiaries	Rs 10,000–25,000 ($200–$500) 1,326 loans of Rs 31.2 million ($624,000)
Repayment	36 instalments	Yearly premium of Rs 100	Single repayment at end of season
Interest rate	12%		16%
Transaction costs	Upfront fee: Rs 20 ($0.40) per loan	Upfront fee: nil	Upfront fee: Rs 20 ($0.40) per loan

A borrower is considered to be in default if he is unable to repay the loan within the 42 months. If the catches are higher than expected, the borrower can pay off the loan earlier without penalty. Over 50% of the fishermen do this, while another 25% pay on time. The remaining 25% are delayed, in which case the repayment will be rescheduled.

The village society pays the amount collected each month to the district federation, which pays it to the SIFFS headquarters. This transfer process enables the SIFFS apex body to approve fresh loans every month, so recycling the funds.

The apex body borrows money from different sources at an average rate of 7% (Figure 6.28, Table 6.11). It charges 10% interest when it lends to the district federation. The interest margin of 3% (10% minus 7%) is among the lowest for a microfinance institute in India. One percent covers the apex body's microfinance operations, and the remaining 2% covers loan losses.

The district federation lends money to the village society at 12% interest. The margin of 2% (12% minus 10%) is used to cover its expenses. The village society also lends to its members at the same rate: 12%. It covers its costs through its 3% levy on fish sales, so does not need to charge extra for loans.

Debt redemption loans

Many fishermen in India are heavily indebted to middlemen who lend cash in return for taking control of the catch. These middlemen charge high rates of commission: 10–15% of the daily catch – however big – and the borrower never has a chance to pay off the principal. The fishermen need cash, so borrow again,

Table 6.11 Sources of loans to SIFFS (cumulative to 2008)

Source	Amount of loan		Interest rate (%)
	(million rupees)	(million $)	
Small Industries Development Bank of India	50.3	1.0	8%
Cordaid	20.3	0.4	7%
Commercial banks	90.9	1.8	7.5–9.5%
Own resources and donated	30.0	0.6	0%
Total	**191.5**	**3.8**	**7%**

driving themselves ever deeper into debt. Such debts can reach Rs 100,000 - a huge sum.

To counter this, the society has developed a "debt redemption loan". The fisherman tells the society how much he owes the middleman. The society pays off this debt. Then each day, it deducts 10% of the value of the catch from the fisherman's account. It uses this money to pay off the amount the fisherman has borrowed, and charges a fixed 12% interest. Typically, a fisherman can pay off the whole debt within 2–3 years (❺ in Figure 70).

Life and disability insurance

Fishing is a risky business. Outboard motors break down at sea, boats capsize, and landing on a beach through heavy surf is perilous. To cover these risks, SIFFS offers insurance to cover instances of death or permanent disability (such as the loss of a hand or leg). It charges a premium of Rs 100 ($2) per year. The scheme will pay out Rs 75,000 ($1,500) in case of accidental death or full disability, or Rs 35,000 ($700) for a natural death or partial disability. This scheme is voluntary, but most SIFFS members have signed up (❻ in Figure 70). Children of people who have signed up to the scheme are eligible for a high school educational scholarship worth Rs 1,200 ($24) per year.

SIFFS's insurance plan is linked to the Janashree Bima Yojana Social Security Scheme managed by the Life Insurance Corporation of India, a government body.

Compulsory savings

In order to inculcate savings habit among the members, a compulsory savings scheme is in operation. Every member contributes 2% of the value of the daily catch as compulsory savings. Members can withdraw their savings for emergencies such as to cover family expenses during the lean season, to repay loans, etc. A minimum balance of Rs 500–3,000 ($10–60) is to be maintained. The society holds these savings and uses them as working capital and to provide emergency assistance to its members. The savings and emergency assistance do not attract any interest (❼ in Figure 6.27). There are no banks or alternative savings institutions in the villages.

Savings for old age security

As SIFFS members became more aware of the need for savings, they started asking the society to provide a savings scheme for old age. SIFFS started such a scheme in 2001. Members can sign up to save a minimum of Rs 50 a month, or multiples of Rs 50 above that. When they retire from the fishing sector, they can get a lump sum consisting of their savings plus interest. The interest rate is announced each year; it is typically half a percent more than the interest rate offered by commercial banks for recurring deposits. If a member delays payments into the scheme, a penalty of Rs 1 for every Rs 50 is levied. The member has the option of withdrawing the amount saved, plus the interest, at the end of every fifth year. This savings programme has been a success. At the end of December 2008, 5,044 members had saved Rs 10 million ($200,000) (❽ in Figure 6.27).

Services for fish vendors

Women's working capital

The women vendors need cash to buy the fish auctioned on the beach, and to pay for transport and other expenses. These women belong to a fisherwomen's society, affiliated with SIFFS, which handles working capital loans for its members.

SIFFS provides loans to the village fisherwomen's societies. The fisherwomen can borrow up to Rs 20,000 ($400) from their village society (❾ in Figure 6.27 and the second column in Table 6.9). These are shorter-term loans: loans up to Rs 10,000 must be repaid within 12 months; larger loans must be repaid within 24 months. The borrower repays the loan every day so the borrower does not have a chance to spend the money on other items. A society collection agent goes round the borrowers' houses in the evenings to pick up the money and takes the money to the village society office. The society office remits the money to SIFFS at the end of each month.

The fisherwomen's society charges 16% interest on these loans. This includes the standard SIFFS 12% interest rate, plus 4% to cover the costs of the society (Figure 6.28).

Seasonal loans

Fishing in south India has "lean seasons" when few fish are caught in certain areas. During these times, the women may travel hundreds of kilometres to buy fish in bulk. They bring this home, salt and dry it so it can be stored, then sell it to bridge them over the lean period. For this they need a larger amount of money. SIFFS grants loans of Rs 30,000 ($600) to cover this expense. The borrower repays

Box 6.9 A retired fisherman reflects

"I joined the society in 1976 and was an active member until last year. But since then, I have not been so active due to ill health.

"For every 5% that the society collects, it retains 3% for expenses, while the remaining 2% goes into our savings. Before the society came into existence, we had no idea at all about savings. Now, our savings depend on our catches – the more the catch, the more are our savings.

"I have taken loans from the society 7 or 8 times. We have to contribute Rs 5 ($0.10) towards a fund to meet expenses related to a member's death or retirement. It doesn't matter where the member dies – if a member in Neerodi dies, a contribution can be expected from members at Manakudi, and vice-versa. There is a lot of unity and cooperation in the community.

"Before the society's formation, whatever price the middlemen fixed was the price we got – in the morning there would be one price, while in the evening there would be another. After the society introduced the auction system, the price has been steadier.

"After joining the society, I could easily get loans for machines and gear. All these have definitely helped me and my family. I have benefited and grown from being a member in the society, and I realize the importance of the SIFFS society for fishermen like me."

Mr Kamalias, Manakudi village, Tamil Nadu

the interest each month, and the principal as a lump sum at the end of the lean period (➓ in Figure 6.27).

Other value chain services

SIFFS provides a wide range of other services to fishing communities in southern India.

- **Boat building and equipment repair** SIFFS owns 15 boat-building and repair yards and 25 outboard motor service centres. These are run on a commercial basis; finance for these comes from a combination of banks and SIFFS's own capital. It is not involved in making nets as this activity is already competitive, so no intervention by SIFFS is required.
- **Research and development** SIFFS develops and promotes improved fishing technologies and ways to improve safety at sea.
- **Fuel supply**. SIFFS cooperatives bulk-buy fuel and sells it to their members.
- **Ice and transport services** Apart from the market facilities and auction system, SIFFS owns three commercial ice plants to supply ice to the villages. It also finances vehicles to transport the fish.
- **Market information** SIFFS provides chain actors with information on market prices at various locations.

> **Box 6.10 Investment loans then and now**
>
> "At first, we were getting loans up to Rs 5,000 ($100) from the society. Now we can get Rs 75,000 ($1,500). We use these loans to buy nets and logs of wood for our *kattumarams* (traditional boats). Earlier, we had to rely on middlemen who gave us loans of 5–10 years duration. But for a loan of, say, Rs 1,000 ($20), even after 10 years we find ourselves still paying the same amount of interest while the middleman retains the original capital."
>
> *Mr Irudhayadhason, Manakudi village, Tamil Nadu*

Accessing finance

SIFFS obtains funds to support its financial services from various sources. It is recognized as a microfinance institution by the Small Industries Development Bank of India, which provides financial support for on-lending to fishing communities. After seeing the success of this on-lending, other commercial banks started providing support.

So far, SIFFS has obtained over Rs 161 million ($3.2 million) of loans from various development finance organizations and commercial sources at varying interest rates (Table 6.11). SIFFS also raises its own money and accepts donations.

As of December 2008, SIFFS had disbursed to its members over 23,000 loans worth a total of Rs 326 million ($6.5 million). Some Rs 97.5 million ($1.95 million) is still outstanding. The average of all types of loan amounted to Rs 18,360 ($367).

Risk analysis

The financial institutions and bankers lending to SIFFS do not demand hard collateral. Rather they rely on SIFFS's strong organizational base and its control over the financial flows in the fish chain. The risks of lending are reduced, because SIFFS has triangular relationships at various levels of the value chain, through which it provides a full package of business services. Here are some examples how the risks are reduced (see Table 6.12):

- The risk of a member not repaying a loan is low because the fish is marketed through the society, and the repayment is deducted right at the source. Default is limited because SIFFS checks how much fish a borrower has caught in the previous year, and because fishermen are fond of the auction system: it guarantees that prices are transparent and the best possible.
- Production risks are limited due to SIFFS's multiple services which include: inputs supplies such as boats, engines, ice and transport, storage facilities, marketing support, and insurance.
- Market risks are low because SIFFS controls the auctions and has direct links with export houses. By pooling their catch and sending them to distant markets if local prices are low, the societies can reduce the risk of local merchants'

cartels. If necessary, the society has storage facilities to store surplus fish (2-ton ice boxes).

- If necessary, SIFFS can use the member's accumulated savings to cover gaps in repayments, during the lean season or in emergencies. Members are obliged to save regularly, which gives them a buffer to draw on. There is a moratorium of up to 6 months to make it easier for the fishermen to repay on time (borrowers must repay 36 monthly instalments within a period of 42 months).

Through these services and relationships, SIFFS has made it possible for banks to lend on the basis of the fishing performance rather than on collateral.

Benefits

Fishermen Instead of being exploited by middlemen and traders, the fishermen can get a good market price for their catch. They are able to access loans at low interest rates (16% as opposed to 36–120%). They can free themselves from permanent debt traps. They can get repeated loans, and can afford to invest in new equipment. They can bridge lean seasons, can save for their old age, and are insured against death or injury. Women can get microcredit to trade in fish or to use for other purposes. Storage facilities have improved with the use of ice and ice boxes, leading to lower losses and more stable prices.

Prices of fish have risen by as much as 90% since the fishermen were organized into societies (Table 6.13). The combination of higher prices and lower interest rates means that the income of each crew member has tripled. For example, in Tharangambadi, returns per crew member rose from under Rs 5,100 ($102) per

Table 6.12 Risk analysis for the fish value chain in India

	Source of risk	Risk level	Mitigation strategy
Production	Lean season	●○○	Repayment holiday
	Migration by fishermen away from area	●●○	Managed by village societies
Price	Oversupply when many fish caught	●○○	Improve storage facilities Fish drying
	Merchant cartels	●●○	Village society bulks produce to sell separately Marketing support
	Low prices for exportable fish	●●○	Contact with export houses to eliminate middlemen
Default	Low fish catches	●○○	Reschedule debt
	Accident	●○○	Loan waiver
	Loss of boat, motor or equipment	●○○	Equipment insurance

Risk levels: ●●● High ●●○ Medium ●○○ Low

year to nearly Rs 18,000 ($360) (Table 6.14).

By organizing in groups to market their catch, the fishermen have been empowered. They have freed themselves from a debt trap and are able to invest in their own futures.

Table 6.13 Fish prices before and after organizing into fishermen's society, Tharangambadi, Tamil Nadu

	Rs	
	Before	After
Seer fish (per kg)	90	170
Tuna (per kg)	15	25
Mackerel (per fish)	6	9
Reef cod (kg)	35	50
Sardines (per 100 fish)	15	35

SIFFS SIFFS has expanded as fishermen and -women have realized the benefits of membership. Membership rose from 3,600 in 2001 to 8,800 in 2008. There are now 153 fishermen's societies and 174 fisherwomen's societies all around the coasts of south India. SIFFS is now an internationally known organization, able to access finance from both national and international commercial sources. Collaboration with the government has improved. Government agencies now want to coordinate their research and extension activities with SIFFS.

SIFFS has increased the range of services it provides to its members. Because it controls the marketing of the product, it is assured of repayment. Default rates are very low: they occur mainly because of accidents. Delayed payments are about 25%, but SIFFS is strong enough to reschedule the debts, charges interest on such loans, and has a way of ensuring repayment.

SIFFS's business activities break even and often generate marginal surpluses; its development activities are based on donor support.

The value chain The value chain is now more efficient because fish is sold in bulk through the societies. Sales are faster, hygiene has improved, and quality has gone up. Links to markets have improved through the organized auction system and direct ties from SIFFS to city wholesalers and exporters.

Competition has been introduced into the chain, forcing other buyers to raise the prices they pay to non-members too. The whole fishing community benefits. SIFFS has its own ice factories, boatyards and outboard repair facilities, forcing other entrepreneurs to improve their services.

Challenges and the future

Loan waivers When the government announces loan waivers in government banks and cooperatives, society members want SIFFS to do the same. This undermines SIFFS's efforts to create a culture of credit repayment. The absence of such a culture meant that other credit institutions refused to provide credit to fishing communities. However, after the 2004 Indian Ocean tsunami, when the commercial banks and cooperatives waived the loans of fishermen fully (both

Table 6.14 Returns to crew members before and after organizing into fishermen's society, Tharangambadi, Tamil Nadu

	Rs		$	
	Before	After	Before	After
Average fish catch	79,412	150,000	1,588	3,000
Loan amount	25,000	25,000	500	500
Interest on loan	9,000	3,000	180	60
Operational expenses	40,000	40,000	800	800
Returns	30,412	107,000	608	2,140
Returns per crew member	5,069	17,833	101	357

principal and interest), SIFFS waived only 6 months' interest and offered 6 months of repayment holiday. SIFFS was successful in collecting 95% of the pre-tsunami loans.

Resource depletion Artisanal fisheries extract a relatively small proportion of the fish in an area, so are sustainable and create a large amount of employment. Large, mechanized boats take a much larger proportion of the catch and create few jobs. They deplete the resource and damage the sea bed. The result is fewer fish for poor artisanal fishermen to catch.

Natural calamities Floods, cyclones, and tsunamis are unpredictable and can be disastrous for fishing communities. They kill people, damage equipment and facilities, and prevent fishermen from going to sea for weeks on end. Global warming is likely to change fish stocks.

New financial products SIFFS is planning to introduce an insurance scheme for boats and motors. Mainstream insurance companies refuse to insure these risks because of mistrust between fishermen and insurance firms. For example, if a boat sinks, there is no way of proving the loss to the company. Closer to the communities, SIFFS is able to control these risks more readily. It is planning to handle individual incidents itself, and to pass on mass claims (for example, due to a cyclone) to a mainstream insurer.

Linking with export houses Prawns, cuttlefish and squid are high-value products with a ready export market in Japan and other countries. Exports are normally arranged by specialized firms that have agents in fishing areas. However, these agents take a large percentage of the value, and often cheat on the weight of the product. SIFFS is developing direct links with the export houses to avoid this problem. This creates a new issue: the export house pays for the produce only after 1 month, tying up a large amount of SIFFS funds. SIFFS is approaching several banks to provide credits to facilitate such exports.

Lessons

Controlling the market By controlling the market, SIFFS has been able to raise the price of fish its members receive. This is possible only because SIFFS is a sufficiently large (or monopoly) player in each village, and has a sufficiently wide geographical coverage that other actors in the chain are forced to trade with it. This market power creates resentment among the merchants, who earn far less if the fishermen in their village organize as a SIFFS-affiliated society. SIFFS staff have been threatened by merchants, and the society has been forced to withdraw from some villages for a time. Eventually, however, the merchants have to become reconciled with their lower incomes.

Local management It is important to design marketing and other services that the local community can manage themselves, after adequate capacity building. The societies are run by fishermen and -women themselves. SIFFS provides them with training in leadership, management and accounting so they are able to manage the village society.

The poor are bankable Small-scale fishermen with highly fluctuating incomes can be good clients for a credit programme if they are organized into a strong and stable organization. The fish marketing programme generates extra income that the society members can save and invest. The poor have multiple financial needs: they need to get out of debt traps, they need working capital, money to invest in assets, insurance, and so on. SIFFS has found a way to provide all these forms of "livelihood finance" by integrating them into a multi-service model and spreading the costs over different services and a larger base. That reduces the risks inherent in providing large sums and reduces the interest rates SIFFS must charge borrowers.

Long-term approach Some development approaches strive to scale up rapidly to reach as many people as possible. The SIFFS model combines institutional and financial development: it is not possible to provide the financial services without a strong local institution. That means rapid expansion is not possible.

Relationship between financial and other processes SIFFS attempts to improve the lives of its members by working on many aspects of the value chain: financial (loans, insurance, etc.), technical (boat-building, motor servicing, ice-making), institutional (the village, district and federal structure of SIFFS, the auction system) and social (the village societies of fishermen and -women). SIFFS began its work with social mobilization to addressing key issues related to fisheries, then moved to finance, marketing and disaster preparedness.

Spreading risk and sources of finance Value chain financing needs to incorporate a variety of institutions, depending on the type of finance, the level of operations, size, risk, etc. Fishing is inherently risky, so the risk of financing is also high. Various savings products for households enable them to cushion borrowers from fluctuating income levels and provide the lender with working capital. During crises, it is important to have a combination of financial products, make a greater provision for risk, and leverage subsidized government schemes.

Action research SIFFS uses an action research model, testing new approaches before rolling them out on a large scale. Keeping the fisheries value chain as the boundary of its operations, it identifies products and services across the value chain that can influence the profitability of small-scale fishing.

More information

*V. **Vivekanandan** vivek@siffs.org or **Ephrem** ephrem@siffs.org, South India Fishermen Federation Societies (SIFFS), www.siffs.org*

*N **Raghunathan** raghu@cms-india.org or **C S Nagabhushana** nagabhushana@cms-india.org, Vrutti Livelihood Resource Centre, Catalyst Group, www.catalysts.org*

7 Synthesis

THIS CHAPTER SUMMARIZES THE main lessons from the case studies and offers ideas for organizations to work on value chain finance as a way to promote rural businesses, both large and small, in ways that are financially successful and that benefit the poor.

Situations of value chain finance

Value chain finance is not always needed. A value chain may work well without any intervention from a financial agent. If value chain finance is not tailored to the capacities and needs of the businesses in the chain, it may even disrupt the value chain, creating bubbles, debts and dependencies. Hence it is important to ask in which situations value chain finance is relevant.

In this book, we have identified three situations where value chain finance has been used:

- **Crafting new chains** Financial services are introduced as part of an integral approach to building a new value chain with smallholder producers. The total intervention package includes organizing farmers, technical assistance, building management capacity, technological upgrading, securing a market outlet, and financing investments or business operations. Our four cases cover milk in India, potatoes in Peru, milk in Bolivia, and chilli pepper in Kenya (Part 4 of this book).
- **Expanding chain liquidity** Financial agents link up with an existing value chain to finance the product flow in the chain, so that the businesses involved can grow and scale up their operations. The main service here is the provision of working capital for the production and trading of the farm product. The cases cover cotton in Tanzania, sal leaf in India, tea in Kenya and rice in Rwanda (Part 5).
- **Unleashing investments in the chain** Financial agents link with an existing chain to provide financial services that enable chain actors to upgrade their businesses through middle-term investments (2–6 years). The cases are honey

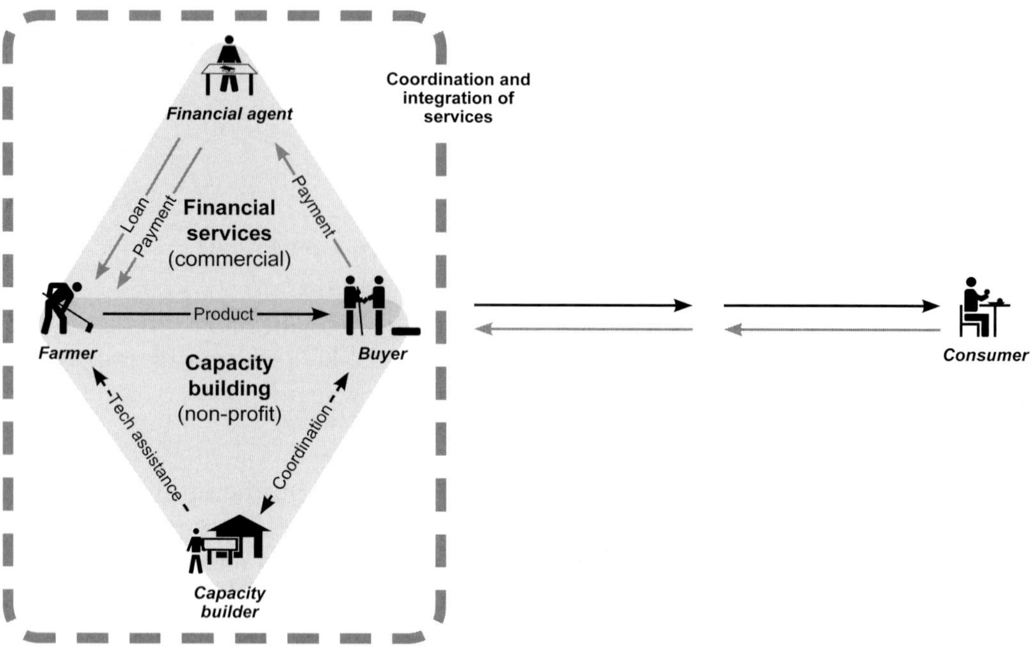

Figure 7.1 Two triangles: Financial services and capacity building

in Kenya, fish in India, soybean in Ethiopia, coffee in Nicaragua and quinoa in Bolivia (Part 6).

We will now look at each of these situations in more detail to see how value chain finance is being used as a way to catalyse rural entrepreneurship.

Crafting new chains

Sometimes a market opportunity exists, but no value chain to supply the products that consumers demand. Or the other way around: sometimes producers have the potential to deliver good products, but there is no value chain to link them to the market. In these situations, it is necessary to build the value chain. Finance is then obviously one of the aspects that needs to be arranged.

Take, for example, our case of milk in India (page 68). The middle class in Indian cities is booming. They want to buy high-quality milk in modern retail outlets, rather than getting their milk from small vendors in the streets. Reliance Group, India's largest industrial holding, sees this as a market opportunity. It sets up a modern milk chain with milk collectors and milk processors who operate under a franchise model. To reach the farmers and integrate them in the new chain, Reliance allies with the BASIX Group, a private company specialized in working with the rural poor. BASIX Group supports the milk farmers with an integrated package of services that are delivered through two separate channels. On the one hand, BASIX's non-profit agent supports the farmers with organizational capacity-building, technical assistance and veterinary services. On the other hand, BASIX's private bank provides loans to the farmers for buying new, more productive milking animals, as well as other financial services such as insurances and saving accounts.

So the new dairy chain in India is built with an integral package of support services that are delivered through two separate triangles of cooperation. One triangle provides capacity-building services to the farmers. This triangle is between the farmers, Reliance and BASIX's non-profit agent. Another triangle provides financial services to the farmers. This one is between the farmers, Reliance and BASIX's commercial bank. The capacity-building triangle is non-profit, while the financial triangle is commercial. For that reason, the two triangles need to be clearly separated into two domains.

We see a similar situation in other cases where an entirely new value chain is built. There are always two triangles (Figure 7.1):

- **Capacity-building triangle** Cooperation between the producer, the buyer and a non-profit service provider. The purpose is to build managerial, organizational and technical capacities among the producers so they become solid chain actors. The cooperation is initially based on non-profit principles.
- **Financial triangle** Cooperation between the producer, the buyer, and a financial agent. The purpose is to provide the producers with access to the financial services they need for their business. The cooperation is based on commercial principles from the start.

The reason that the triangles are separated is that one is non-profit while the other is commercial. Investing in the basic capacities of producers cannot be done on a commercial basis, at least not initially. But financial services must be commercial from the start, otherwise you create a corrupted culture of non-payment and non-compliance. The producers need to have a stake in the business: they need to be rewarded for good performance, but also feel pain when they do not comply.

The fact that the triangles are separated does not mean that they are not interrelated. On the contrary, there must be close coordination to ensure that the financial services are tailored to the capacities and needs of the producers. Such coordination can occur at two levels. One is at a high level between the organizations involved, for example through memoranda of understanding as used by the K-REP group for the honey chain in Kenya (Chapter 6). The other is through the fieldworkers who are involved in the capacity building triangle. They can indicate precisely when the farmers need money for buying fertilizer, hiring workers, etc. They can also indicate how much money the farmers need for investments in equipment, animals, etc. Through their frequent farm visits, the fieldworkers can monitor whether the loans are used in a proper way and whether the farmer has the capacity to repay the loan. In this way, the capacity-building triangle provides soft collateral for the financial triangle. It is the combination of the two triangles that makes value chain finance viable in the high-risk start-up phase of the new chain.

But what happens when there are no more funds to pay for the capacity-building triangle? The major risk of crafting new chains is the creation of lasting dependencies on the service-provider. Too often we see that the entire value chain collapses after the withdrawal of the service-provider (for further information and examples see KIT, IIRR and Faida Mali, 2006). Our case of potato in Peru (page 40) provides some valuable insights to prevent such disaster.

In Peru, Snacks, a Pepsico subsidiary, produces crisps from a special type of potato. However, Snacks was short of supply of this potato, and decided to ally with the NGO Fovida to get more farmers in the value chain. In the beginning, to get the chain going, Fovida assumed a role as an intermediary in the chain. It organized the farmers, provided technical assistance, arranged for inputs, bought the potatoes and delivered these to Snacks. Fovida also arranged for farmers to get financial services through its sister microfinance institution, Credivida. So there was a capacity-building triangle and a financial triangle.

Initially the services from Fovida were paid for by a donor, but Fovida was aware that this would be unsustainable on the long term. So after 3 years of operation, when the chain was working well, Fovida agreed with the farmers on a service fee of 1.27% of the value of the potatoes. After another 3 years, when donor funds were no longer available, the service fee was doubled to 2.53%. The farmers were willing to pay because they saw the value of Fovida's services and they were well informed about the costs of these. Also, Fovida did not oblige the farmers to use its services. Farmers were free to buy inputs and technical services elsewhere, and they could use the credit history that they had built to get loans from other parties than Credivida.

Now after 8 years, Fovida has phased out completely. The chain is self-sustaining. There is no more need for triangles. The farmers now have a direct contract with Snacks, and they can choose the technical and financial service providers of their own preference.

The Peru case provides some important insights for avoiding lasting dependencies:

- **Do not start without an exit** Plan the exit strategy from the beginning, and inform all chain actors about it. In the course of the project, build the capacities of the chain actors so that they will be prepared for take-over. From the start, be transparent on the costs of the services you deliver and enable the farmers to choose to use your services, based on knowledge of the costs.
- **A value chain is business, not a project** The chain has to be market-driven. You start with detecting the market demand and screening, selecting and involving the relevant business partners. Only after that can farmers and community groups become involved. Make sure that all actors that you involve are prepared to take risks, because involvement in a chain is business. Make sure that the chain creates added value for all actors, so it is sustainable in the long term.

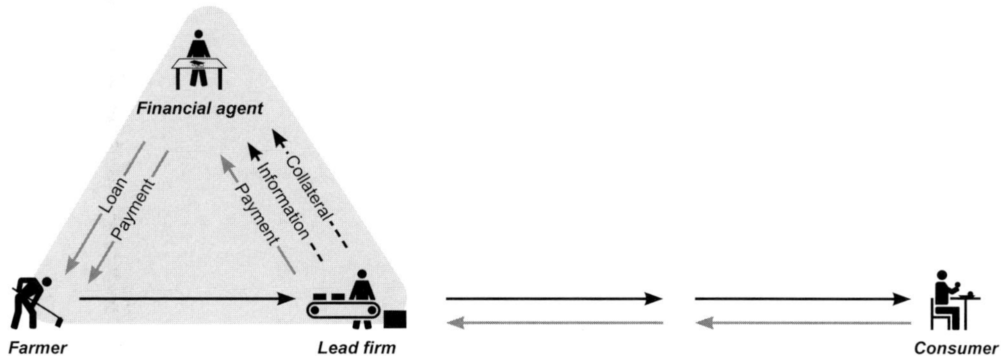

Figure 7.2 The lead firm strategy

- **Take your time** The potato chain took 8 years to build. This is more time than most donor projects allow. So take your time and once you start, be prepared to be in it for a long time.
- **Building and destroying triangles** In the Peru case the relationships changed continuously. Fovida began as a chain actor, evolved into a service-provider, and then phased out. First the two triangles of finance and capacity-building were put in place, but later they were replaced by direct bilateral business relations.

In sum, when a new value chain is crafted, there need to be two triangles of cooperation: one non-profit triangle for capacity-building, and one commercial triangle for financial services. They need to be separated but well coordinated. The capacity-building triangle forms soft collateral for the finance triangle. In the course of time, as the producers develop and the chain consolidates, the triangles may disappear. When the triangles are fully gone, you have reached a situation where the producers are fully empowered to have direct business relations with buyer and financers.

Expanding chain liquidity

Many value chains work reasonably well, but they cannot realize their full potential due to a lack of working capital. For many traders, farmers and small-scale processors, working capital is the main limiting factor for further business growth. They have no collateral and therefore no access to bank loans. How can financial agents start to work with these enterprises? How to finance entrepreneurs who do not have sufficient hard collateral?

The cases in this book show two different basic strategies:

- **Lead firm strategy** When there is a strong lead firm in the chain, this company can be used as a vector to reach non-bankable suppliers. The lead firm is the collateral for providing finance to the suppliers. So the financial agent establishes a triangular cooperation with the lead firm and the suppliers. This

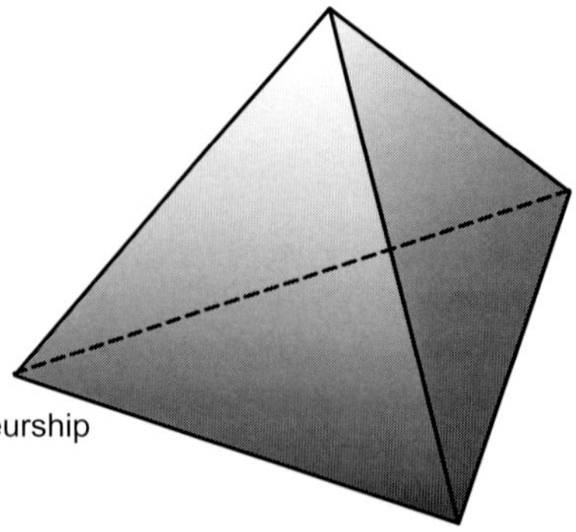

Figure 7.3 Three basic elements for the capacity-building strategy to make small-scale suppliers bankable

we see in the cases of tea in Kenya (page 125) and organic cotton in Tanzania (page 94).

- **Capacity-building strategy** Another strategy is to build the capacities of the suppliers to make them bankable. When the suppliers are trained and supported in their businesses for a couple of years, they can build a proven credit history and then become eligible for financial services from mainstream banks. This strategy has been applied in the cases of sal leaf in India (page 110) and rice in Rwanda (page 135).

We will now take a look at both strategies in some more detail.

Lead firm strategy In the lead firm strategy, a triangle is built between the producer, the buyer and the financial agent (Figure 7.2). The loan goes to the producer, while the buyer pays back the loan. So the due diligence for the loan is performed on the buyer. It is irrelevant whether the producer can provide collateral. When the buyer is a solid, respectable lead firm with strong linkages to the farmer, this strategy is very powerful in delivering financial services to small-scale farmers and suppliers.

For example, in the Kenya tea case (page 125), the farmers receive cash payments for the tea that they deliver to the tea factory through the services of a factoring house. The farmers deliver tea to the factory, issue an invoice, and the factoring house pays the farmers up to 70% of the value of that invoice. After 30–60 days the tea factory pays the factoring house the full value of the invoice, and then the farmers receive the remaining balance. In this case, the factoring house performs its due diligence on the buyer – the tea factory. The buyer is the collateral for the factoring service, as the buyer is the one who will pay back the loan. The seller – that is, the tea farmer – is not important for repayment. So there is no need for the tea farmer to offer collateral. The key criterion here is that the buyer should be a respectable lead firm.

The same occurs in the case of organic cotton in Tanzania (page 94). Tanzanian cotton exporter BioRe receives an international working capital loan from Triodos Bank, so it can pay its farmers on the spot for their seed cotton. BioRe's balance sheet is insufficient to secure this multi-million loan. The guarantor for the loan is BioRe's client Remei, a trader based in Switzerland. The loan is collateralized by the export contract and the intimate business relationship between BioRe and Remei. The fact that they jointly set up the cotton chain, integrating all links from farmers to ginner, spinner, textile manufacturer and final retail, is the soft collateral that gives Triodos Bank the confidence that the loan will be repaid. The trust, transparency and vertical coordination in BioRe's cotton chain underpins the loan without further need for hard collateral.

Capacity-building strategy The capacity-building strategy focuses directly on the suppliers. It aims to build the capacities of the suppliers to make them bankable. When the suppliers are trained and supported in their businesses for a couple of years, they build a proven credit history and then become eligible for financial services from mainstream banks. This strategy requires three basic elements (Figure 7.3):

- **Entrepreneurship building** Each individual supplier must be trained to build capacities in financial and business management. The supplier must evolve into a skilful entrepreneur, otherwise he or she will never become bankable.
- **Organization building** Small-scale suppliers who are isolated will most likely never become viable chain actors. The suppliers need to be grouped in a strong organization so they can develop effective business relations in the value chain and with financial agents. Through their organization the suppliers are empowered as a value chain actor. The organization also serves as a mechanism to deliver finance to the suppliers, and the organization can evolve to become guarantor for loans.
- **Value chain building** In many situations there are also bottlenecks in other segments of the value chain. Bottlenecks may occur in chain functions such as inputs, trading, transport or processing. Or bottlenecks may occur in chain management such as quality assurance, innovation or information. Such bottlenecks need to be resolved for the chain to reach its full potential.

For example, in the case of sal leaf in India (page 110), large numbers of small collectors, processors and traders make a living from the production of plates from sal leaves collected in the forest. Their businesses remain very small due to a lack of working capital. A problem is that the small-scale suppliers are not perceived as bankable. Generally, loans from banks and microfinance institutions are too large, too expensive or too burdensome to get. The international agency UNIDO intervened at two levels. One, it organized the collectors and traders and built their capacities in accounting. This was necessary as these small-scale suppliers lacked business skills. Second, it worked with the banks to create understanding and redesign their financial products to the needs of the sal leaf cluster. In this way UNIDO contributed to value chain building through providing banks with information on the sal leaf cluster. Now the micro-entrepreneurs in the sal leaf cluster have access to financial services that suit their needs.

Another example of a capacity building strategy involves the rice case in Rwanda (page 135). Rice trade in Rwanda was dominated by Chinese entrepreneurs. To defend themselves against the middlemen, rice farmers formed a cooperative, COPRORIZ. Initially the cooperative faced some severe financial problems. Due to a lack of working capital the co-op was not able to distribute inputs to their members; furthermore, it lacked cash to pay farmers directly and was making a loss. The microfinance institution CAF Isonga intervened and has become the main source of finance for COPRORIZ. CAF Isonga built the entrepreneurial capacities of the farmers by making it possible for registered farmers to be paid cash on delivery through a voucher system. CAF Isonga also contributed to organizational building of the co-op. CAF Isonga started to provide credit to farmers, enabling the co-op to provide their members with inputs. Membership of the co-op is the collateral for getting a loan. To make sure that CAF Isonga can play its role CAF Isonga is being informed on the product flows entering and leaving the rice warehouse of the coop.

Capacity-building strategies go hand-in-hand with making available (new) financial products that meet the needs of the involved suppliers. These products can expand the chain liquidity of the involved businesses. Expanding chain liquidity is a powerful way to boost the value chain and enable the chain actors to grow and scale up their businesses. However, it should be kept in mind that it is feasible only in value chains that show a healthy balance between supply and demand:

- On the one hand, there needs to be a strong market demand for the produce, otherwise it is unclear whether there will be revenues to repay the loan. If the financer is not convinced that there is a market for the farm produce, it will not proceed to issue the loan.
- On the other hand, market demand should not be too strong, otherwise there is the risk that the farm produce will be diverted to competing buyers who offer a slightly higher price. The chain needs to comprise mechanisms that assure that the produce will be actually delivered to the guarantor of the loan. For example, though a solid contractual relation between supplier and lead firm, or through organizational discipline with a farmer group.

In conclusion, the strength of the value chain determines its bankability. When the chain actors work closely together, they have better access to financial markets. Working together with a strong lead firm can help in accessing finance. Cooperation between chain actors is the soft collateral for obtaining financial services. Therefore, it is important that all chain actors are visible and have entrepreneurial skills, which requires building their capacity.

Unleashing investments in the chain

Besides the lack of working capital, another key reason value chains may not realize their full potential is the lack of investment capital. Entrepreneurs need investment capital to upgrade their technologies, introduce new products, develop new markets, etc. The third situation where we have found value chain finance is when financial agents link up with the value chain to enable the chain actors to make medium-term investments (2–6 years). This is probably even more challenging than providing working capital, due to the risks inherent in medium-term financing. How do financial agents manage these risks? How can investments in the value chain be financed in a viable way?

In the cases we have encountered various examples of medium-term financing, using different investment finance products. But how do these different products promote the business?

Medium-term loans In two cases (coffee in Nicaragua, fish in India), medium-term loans were provided to suppliers. Because of their size and conditions, these credits make it possible to make investments in the chain.

In the case of organic coffee in Nicaragua (page 174), FDL (a microfinance institution) provided a 4-year loan to small-scale coffee growers. The money was paid to the farmers via their union, UCPCO. This needs to approve the farmers who receive the loans. In return, the farmers agree to deliver their coffee to the union. On delivery the farmers get paid half of the coffee price immediately. The rest is paid to the farmer after deducting an instalment of the loan, plus interest (14%). Union technicians pay regular visits to the farmers and reports relevant information to FDL. Farmers who want to borrow a relatively small amount (less than $1,500) organize in "solidarity groups". These groups are responsible for ensuring that the individual members pay back the loans. In case of individual larger loans, farmers need hard collateral, for example livestock. The credits have made it possible to invest in post-harvest processing facilities at the farmers' and the union levels. These enable the producers to comply with the high standards in the market for organic coffee.

In the case of fish in India (page 200), medium-term loans were provided to suppliers of fish who are organized in village societies. Fishing entails high material costs, as boats can leak, engines break down, etc. The village society provides loans to its fishermen members to cover these costs. Members are allowed to borrow up to 25% of the value of their previous year's catch. The pay back time is 42 months, and starts immediately after the money has been borrowed. The village society deducts 10% of the daily catch, and calculates the balance each month. Interest is calculated on the balance of the loan remaining. The society pays the district federation, which in turn pays SIFFS (which acts as a microfinance institution). This transfer process enables SIFFS to recycle the funds and offer new loans. The fishermen have gained enormously, as they are no longer dependent on unscrupulous traders and moneylenders. In addition, they have built their organizational capacities, developed market channels and built up a credit history. Now they are becoming eligible for loans from the mainstream banking sector.

Micro-leasing In the honey case in Kenya and soybean in Ethiopia micro-leasing is provided as an innovative finance investment product. In Kenya (page 149) micro-leasing provided by the K-Rep Development Agency enables beekeepers to lease and maintain expensive hives and increase the quality and quantity of honey they produce. To get a lease, a beekeeper has to get two quotations from suppliers of hives and equipment, and submit these to the village financial services association. The association approves of the request and chooses a supplier, to whom it pays 60% of the price. After delivery (within 3 weeks) the supplier gets the rest of the money. The beekeeper pays off the lease each month over 2 years. Once the beekeeper has paid all the instalments, he or she becomes the owner of the hive. To qualify for this arrangement, the beekeeper must belong to a group and attend training on hive management. The lessee has to report to the group of beekeepers periodically. The group, which has also received training on financial management, takes responsibility for the equipment.

In Ethiopia (page 164), micro-leasing was provided as a service to women's groups for leasing a soybean-processing machine. Buying this expensive equipment was perceived as too risky, so Harbu, a microfinance institution, proposed to lease it to the women over a 5-year period, charging 10% interest a year. This arrangement was based on the long-term trust relationships between Harbu, the women's groups and the farmers' marketing associations. Harbu remains the owner of the machine until the women pay it off fully. Because of the machine the women have been able to scale up their business and increase their income.

Temporary equity Financial investment can also involve the provision of (temporary) equity, which we see in the case of organic quinoa in Bolivia (page 187). When Irupana decided to focus on the processing and export of organic quinoa, it could not convince local banks to provide it with loans. So it approached Pro-rural, a financial institution that specializes in small businesses. The two organizations agreed to a temporary joint venture lasting 2 years. The first joint venture saw Pro-Rural investing $150,000 and Irupana about $100,000 in the marketing of quinoa. This venture was a success: it achieved a return on investment of 8.7%. In a second joint venture lasting 3 years, Irupana invested the larger share, $272,000 (52%), while Pro-rural had a minority of the shares (48%) and so reduced its risk. The venture was also successful, yielding a return on investment of 8.1%. This kind of arrangement has several benefits: it shares risk, the business is transparent to both parties, and both parties have a say in the business. Pro-rural has used a similar approach in its temporary joint ventures with traders.

These various investment finance products manage risks in different ways. We can distinguish four methods of risk mitigation:

Group formation and building the organizational capacity of suppliers Membership of a group is often set as a condition for investment loans or leasing. In the Indian case, fishermen had to be member of a village society. Beekeepers in Kenya also had to be part of a group to get a leasing contract. The groups are responsible for repaying the loan and, in case of leasing, for maintaining the equipment. In Kenya the beekeepers received training on hive management and financial management. Investment in marketing groups also strengthens the capacity of suppliers and contributes to their empowerment. Strong supplier groups benefit the chain as a whole and reduce the risks for financial institutions.

Combined investments with other financial and non-financial services In most cases medium-term investments were combined with other financial and non-financial services. For example in India, SIFFS provides various types of financial services, such as old-age security savings and insurance, together with a number of non-financial services, such as boat building and repair, market information and transport. These multiple services show how risk management in value chain finance can occur. In the case of honey in Kenya, the combination of different financial (insurance and savings) and non-financial services (technical services) considerably reduced the risks involved in honey production.

"Ownership" in the hands of financial institutions This way of risk mitigation is characteristic of leasing arrangements. In the case of honey in Kenya, micro-leasing reduces the risks for the financial institution, since it remains the owner of the hive until the beekeeper has paid it off fully. Micro-leasing in Ethiopia worked in a similar way: the microfinance institution leased the machine to the women soybean processors over a 5-year period but remains the owner until the women pay it off fully.

Risk sharing Another way of mitigating risk is by sharing risks and becoming directly involved in the activity. An example of this is the case of organic quinoa in Bolivia, where both Irupana and Pro-rural invested in a joint venture. They combined their expertise and resources to develop a particular business. Risks can also be shared by working with a variety of financial agents, as in the case of coffee in Nicaragua.

All in all, we have found considerable innovation in financial products for investments in the value chain. These products include medium-term loans, equity and leasing. Various strategies are used to reduce the risks, including strengthening of groups of borrowers, parallel support services to borrowers, and risk-sharing among various actors.

Nevertheless, it is obvious that unleashing investments in the chain is still the least developed form of value chain finance. The investment-capital needs are beyond the scope of the financial products encountered in our cases. There remains a great challenge to enable chain actors to make medium-term investments in their businesses.

New models of risk management

Traditional finance fails to address the needs of small-scale entrepreneurs because their needs are considered too risky to finance. The essence of value chain finance is the creation of new models of risk management which enable better services to small-scale entrepreneurs. Three issues are fundamental in these new models: bringing the business into the picture, the use of the value chain as soft collateral, and looking at the future rather than the past.

Bringing the business into the picture

Rather than looking at the balance sheet and profit-and-loss statements, the financial agent looks at the business case: at the crop, buyer, market, technical assistance, and value chain relations. The financial product is closely linked to the production and trading of an agricultural product within a specific business relationship. So the risk calculation is no longer confined to the standard financial ratios of solvency, liquidity, etc., but rather to the solidity of the business case itself. For example, in the case of organic cotton in Tanzania, a risk analysis of the value chain, rather than an analysis of the financial sheets of BioRe, convinced Triodos Bank to finance this business, looking at market conditions, production conditions, competition from other buyers, etc.

Value chain as soft collateral

By bringing the business in the picture, value chain finance resort to other forms of collateral. It partially replaces traditional hard collateral with the soft collateral that is inherent in the business case. Soft collateral offers proven risk reduction strategies: market risks are reduced through sales contracts, production risks through technical assistance, management risks through producer group formation, moral risks through regular information and communication, repayment risks through a claim on the product.

The cases provide various examples of soft collateral. For example micro-leasing puts a claim on the product or processing equipment, as in the cases of soybean in Ethiopia and honey in Kenya. In a number of cases, group formation helped to manage the business, for example for the fish case in India. In all cases, existing relationships between different actors in the chain provided a basis for trust for the financial institution.

Future rather than the past

By looking at the business case rather than the financial records, value chain finance takes a perspective fundamentally different from traditional finance. Traditional finance looks at the past, as represented by the financial records. Value chain finance looks at the future, at the potential of the business case.

Value chain finance will only work if financial agents learn to build their risk assessment on the basis of the business case. They must develop expertise on the volatile nature of agriculture, the specific commodity sector in question, the workings of value chains, and the functioning of producer groups. This information need is crucial for a financial institution to step in. If the right information is available this in itself is a risk-reducing mechanism.

The chain of value chain finance

Value chain finance is all about collaboration. There is collaboration between the financial agent and the chain actors and service providers (the triangle). But often there is also collaboration between various financial agents working together to serve the value chain: local banks, international banks, microfinance institutions, development finance organizations, insurance companies, and credit and saving associations. This is what we call the chain of value chain finance (Figure 3.4 on page 33).

The chain of value chain finance further reduces the risks and transaction costs of financing the value chain. We have seen three specific purposes of cooperation among financial agents: covering all needs in the value chain, sharing the risks, and reaching out to producers.

Covering all needs in the value chain

Financial agents work together to attend all the financial needs at various levels of the value chain. For example in the case of sal leaf in India, the Ahari Federation provided production loans to organized sal leaf collectors to develop their businesses. Through the Betnoti Federation, banks provided loans for working capital. In addition, banks started to finance some traders.

In the case of honey in Kenya, we also see a chain of value chain finance. The K-Rep group developed into an umbrella organization involving different units targeting different levels of the chain. For example, K-Rep Fedha offers simple financial services (credit, savings) to the beekeepers through village-based cooperatives. Traders have access to short-term loans for working capital and micro-leasing of equipment. The K-Rep Bank provides a new service, factoring, to processors, enabling them to buy crude honey from the collection centres, and to sell semi-processed honey. This combination of arrangements helps smooth the chain.

Sharing the risks

Financial agents work together and share the risks of a specific financial product. For example, in the case of coffee in Nicaragua, initially three sources of finance were used. The farmers' union provided loans to its members as working capital. The microfinance institution, FDL, provided short-term loans to individual farmers. Coffee growers in need of immediate cash can also borrow from private traders. Later a governmental rural development agency (IDR) stepped in to assist farmers in upgrading their processing equipment, by providing them the capital that covered 70% of the total costs involved. FDL also expanded its services. Besides working capital, it also started to provide farmers with investment capital. Furthermore, FDL provided the farmers' union with investment loans and working capital. For FDL it was central to reduce its risks by building on a strong alliance with farmers and the farmers' union.

Reaching out to producers

Financial agents work together with a community-based financial agent (for example a savings co-op, local bank or microfinance institution) to reach out and build interpersonal relations with the credit users. This reduces transaction costs in money flows and ensures proper due diligence and monitoring. For example in the sal leaf India case, two federations of self-help groups were set up. These self-help groups each contributed to a common fund, which was used to get a matching grant from UNIDO. The groups can access credit based on their own business plans. The groups assist their members by supporting their loan requests and pressuring them to repay on time.

In the case of honey in Kenya, the village-based cooperatives (financial services associations) were owned by their users. Each association had about 2,000 members (including beekeepers, farmers and other local people) who owned shares in it. These associations were able to offer tailor-made services to their members. The borrowers did not need hard collateral; the groups took the responsibility for paying back the loan.

In this book we have seen chains of value chain finance in six case studies: honey in Kenya, rice in Rwanda, soybean in Ethiopia, and fish, milk and sal leaf in India. This is no coincidence. Apparently an integrated approach in value chain finance works better – it is not possible to strengthen the value chain unless all the links work well. The importance of the chain of value chain finance shows that effective intra-organizational cooperation is needed to enhance small-scale entrepreneurs' access to finance. This is not an area where individual organizations can resolve all challenges. Microfinance institutions can achieve a lot, as can individual banks and trade finance funds. But the needs of small-scale entrepreneurs in value chains are only tackled effectively if financial agents work together and different services are provided.

Involving local banks

Local banks need to play a key role in enhancing the access to finance for small-scale entrepreneurs. As stated above, microfinance institutions and international banks can achieve a lot, especially in the pioneering phase, but to mainstream value chain finance, local banks need to get involved. How is it possible to stimulate local banks to provide working capital to value chains? In the cases we have seen several elements that are important, as described below.

Making the chain actors visible

It is necessary for the actors to be seen as solid, bankable companies. In this respect it is important to ensure that the actors are registered as legal entities, with audited books. Besides available information and transparency, some level of organization among the suppliers helps. For example the sal leaf collectors and plate makers include some of the poorest people in India. These (mainly) women from tribal families were encouraged to form groups and to start saving. At a later stage UNIDO stepped in to support these groups in accounting, preparing business plans, etc. With the training they have received, they are able to write their own business plans and access credit.

In the Peru case, too, organizing the potato farmers resulted in an agreement with Snacks, a market leader in the market for crisps, with the NGO Fovida acting as intermediary.

Developing collateral in the chain

The buyer's guarantee is crucial to convince banks to lend money. In value chain finance, despite the lack of hard collateral, the presence of soft collateral can convince financial institutions to provide their services to rural entrepreneurs. Soft collateral can take different forms: for example, the working relations between different chain actors, or assigning to the group the responsibility for repaying loans. Savings are key in this respect, since they demonstrate the group's financial capacity (this is shown in many of the cases). Other examples of soft collateral include the leasing of equipment, and organic certification (which reflects improved management practices and solid chain relations).

Co-financing schemes

Collaboration with other financial institutions can convince banks to provide additional funding. This was the case in the coffee case in Nicaragua. Due to the growth in the value chain and the good reputation of the union of coffee producers, three foreign development-finance institutions (Root Capital, Rabobank Foundation, and Shared Interest) started to give loans to the union. The union uses these loans as working capital and for investments. The loans are relatively favourable for the union, which no longer needs working capital loans from FDL, the microfinance institution that was involved in building its capacity. There is potential, and need, for local banks also to become involved in such chains.

Piloting

Banks do not always understand how things might work. In this case it is important to start with a pilot to show the opportunity and how to manage risks. The sal leaf case from India gives an example of such a pilot. UNIDO trained the self-help groups on financial practices and management. It also sensitized the banks on the credit needs of the sal leaf collectors who formed the group, and explained how the value chain works. For this purpose different workshops were organized. After a small, successful pilot, the banks were convinced of the bankability of the collectors. They began lending money to the processors to upgrade their equipment, which enabled the process to be scaled up. UNIDO convinced bankers to invest in traders in a similar way. Traders had no accounts or balance sheets, but when they had been trained to improve their financial records, the banks became confident and started to lend money to them.

Helping banks build commodity sector knowledge

Information about the chain is crucial to convince bankers to become involved in value chain finance. To build this knowledge, banks should develop partnerships with specialist organizations (government agencies, companies, NGOs) and establish strong links with actors in the value chain.

Showing the opportunity

Banks may not realize that financing a value chain can be good business. It may be necessary to show them the opportunity: when the banks see a profit, they normally jump in. The financing of chain liquidity, in particular, is a clear market opportunity, because it is possible to charge relatively high interest rates. Production and trade are completely different in their speed and return on capital, so the interest rates should be different, but in trading, high interest rates are not necessarily a problem.

Price is not the only determinant of the quality of the loan. What matters is that they get a loan at the right time. If producers are willing to sell their crop now at depressed prices, rather than waiting and get high prices in 3 months' time, they are in effect buying a very expensive loan. The coffee growers in Nicaragua are just one of the examples in the cases of producers selling their product to private traders in return for immediate payment at low prices.

Leasing is also a profitable opportunity for bankers, with collateral and high interest rates. The two examples of micro-leasing (soybeans in Ethiopia and honey in Kenya) show the feasibility of leasing for both the chain actors and the financial institutions.

Building a track record

In the end, the most powerful way to get local banks involved in value chain finance is through numbers on paper. This means building a track record with the value chain actors. In many cases, we see that after a couple of years the chain actors have gained good access to mainstream banks. Outside organizations can help in building such a track record, as occurred in the case of sal leaf in India (where UNIDO helped collectors and traders), and in Peru (where Credivida assisted the potato growers).

Role of subsidies

The role of subsidies should be carefully considered, especially because value chain finance involves financial and non-financial services. In the discussion on crafting new chains, we said that financial services must be commercial from the start. A number of issues should be taken into consideration before subsidies are provided.

Need for a good diagnosis

There is a tendency to think that poor people need money, but this is not always the case. More than money, they may need other assets to help them improve their business. For example in the case of sal leaf in India, the Dumapada self-help group involved in collecting leaves and making plates needed training and information (Box 5.2, page 115). When they had a loan, members did not know what to use it for. After financial counselling they developed a business plan, decided that the loan was too high, and returned what they did not need to the bank.

Subsidies only at the start

When crafting a new chain, subsidies to reduce risks may be appropriate, for example for training, capacity building, improvements to facilities, or adding value at critical points in the chain. At later stages it is important that farmers start to pay for the services and realize their value. In the potato case in Peru, for example, Fovida gradually increased its service fee. In addition, Fovida does not oblige the farmers to use its services: farmers are free to buy inputs and technical services elsewhere. It is important to focus subsidies on intangible goods and services, such as technical assistance, group formation, training, etc.

Subsidies can also be helpful in case of new product development and for studying risks, for example in the case of insurance. It is good to keep in mind that NGOs and many other chain promoters are almost always subsidized - so cannot continue providing their services indefinitely.

Subsidies for infrastructure only as co-funding, and only when farmers are ready

Subsidies can be used to co-fund chain infrastructure, but be careful: farmers have to be ready. In case of milk in India, the government invested a lot in the infrastructure to support the diary industry, for example by establishing diaries. But these diaries remained underused. The lesson was that farmers do not take the responsibility over a diary if it is not theirs. Reliance Diary Foods Ltd recognized this problem and started to invest in diaries, but this time using a franchise model. They trained local people to run the diary as a business.

There are two other important criteria before subsidizing infrastructure. First, subsidies should be handed out only as matching funds, never 100%. Second, they should only be given in a start-up situation, when there is a need to experiment with new products and new technologies.

No subsidies for financial services

Lending money (such as debts, capital funds) should not be subsidized; only the initial equity or set-up costs could be co-funded. Subsidizing financial services creates unfair competition and impedes private banks from becoming involved. In many countries there is a history of governmental development banks and rural development projects that have used subsidized interest rates. This has had a tremendous effect on the willingness of private banks to involve themselves in agricultural finance. Moreover, it often led to corruption and contributed to poor risk assessment. Loan approvals were often related to political ties and friendship relations instead of business criteria and risks. Resources for subsidies are usually limited, so there is a need for rationing. This can be done with transparent criteria, but often it is done in an informal manner that can easily be subject to fraud.

Subsidies should be temporary, transparent, and preferably free of government interference. A no-payment culture must be avoided. Such a culture can occur if people are pressed to take a loan, as happened in the sal leaf case in India. Before the chain was improved, some borrowers did not even realize that they had gone into debt.

Transparency on costs, and having an exit strategy

If non-financial services are subsidized at first, the recipients of these services should be made aware of the value (or costs) of these services. This is important if after some time the service-provider starts to ask them to pay for its services. Phasing out, as we have seen in the potato case in Peru, requires transparency on the costs and also, from the beginning, on the plans to phase out. The other chain actors have to be prepared for this.

Enabling environment and governance

A value chain does not operate in isolation. Its operations may be supported or hindered by other actors and institutions surrounding it: the national government, local authorities, political interests, powerful competing companies, ethnic tensions, local elites, and so on. It is very difficult to start and sustain a value chain in the face of determined resistance from powerful parties, or if the environment is not enabling. Patronage may favour certain actors and damage the prospects of others. One actor in the chain may become dominant and start to exploit the others, reducing their value shares, restricting their access to credit, and imposing ever more stringent requirements on them.

Nevertheless, value chain finance may also empower weaker actors. Farmers improve their skills in management and marketing, and gain a voice in the chain. They can apply these capacities to seek other markets or expand their production of other types of produce. Value chain finance may give them the step up needed to expand beyond a single value chain.

Various levels of chain governance are important in the cases, including producers' organizations and the national and local governments.

Organizing the suppliers

At the local level, producers' organizations are very important. Organizing suppliers helps them to negotiate, get information, achieve greater efficiency through higher volumes and lower transaction costs, and obtain credit. From another perspective, organizing suppliers helps other actors reach the poor and use social pressure within groups to ensure compliance with quality requirements, delivery criteria and repayment obligations. In the case of rice in Rwanda, we can see how use was made of social structures that already existed. A cooperative played the role of credit officer: it pressured its members to repay.

However, not all existing structures necessarily work. For many suppliers, cooperatives still have negative political connotations, and may be managed badly. This makes good leadership and training of the farmers very important. What is key is that farmers start working together despite these problems.

Role of the government

The role of the government differs from country to country and from one economic sector to another. It may have a facilitating role, or it may interfere or obstruct efforts to develop value chains. Without a good understanding of the role of national and local governments in value chain finance, opportunities may be overlooked or the effectiveness of services that are provided may be hampered.

In some countries progress very much depends on who is in power. When vested interests become involved, progress can become difficult. This goes for every continent, every country and every locality. Progress can also become difficult when a government or local administrator changes, as then everything else may change. This is a frequent problem in Latin America. In some countries, the government plays an overly strong role in certain sectors, or in the economy as a whole. The government controls the land and resources, and determines who has access to these assets. This situation is prevalent in a number of African countries. Banking regulations can have a negative impact on efforts to develop value chain finance. Banks may be sanctioned by the central bank if they offer loans that are not secured with hard collateral (e.g., by requiring them to keep high amounts of unproductive reserves, or to make deposits for them at the central bank.

Involving the government and understanding its role can be a necessity but can also be an opportunity, as projects can be very costly. Governments can also play a role in mobilizing people by showing them business opportunities.

Involving all stakeholders

Building a chain involves regular meetings with all actors, and sharing information on all interests and possibilities. Each stakeholder must be willing to recognize and respect its own role and that of the other chain actors. In the case of sal leaf in India, the creation of platforms facilitated regular meetings among different actors. Such mechanisms are especially valuable where information flows are impeded, as where a market is far off.

Opportunities and limitations of value chain finance

Value chain finance is an empowerment tool for the poor, but at the same time a profitable market opportunity for the finance industry. Here we summarize some of the main opportunities and limitations of value chain finance.

Including the poor

For many poor people living in rural areas, accessing finance is difficult. A value chain finance triangle is necessary to provide them with the funds they need. Nevertheless, for the poorest access to value chain finance will remain difficult as they are often not structurally involved in a supply chain. For value chains to include the poor, initiatives should also ensure that they are empowered by building their capacity and involving them in decision-making. Value chain finance should favour the poor, counterbalancing the monopoly power of strong chain actors. In many of the cases, value chain finance brings benefits to the individual actors and to the chain as a whole. But it is necessary to guard against allowing monopoly power to build up at one key point in the chain. A number of the cases are built around a monopoly: for example, TARDA in the honey chain in Kenya, or Snacks in the Peru potato case.

So value chain finance may result in a paradox: a situation where empowering the poor is a goal, but the finance arrangements are based on the credibility of a single, lead firm. This implies that the poor should be linked to these lead firms. Can a monopoly go hand-in-hand with empowerment of the poor?

One option for farmers and other suppliers is to move away from the lead firm by diversifying markets and products. Another option is to move towards the lead firm, by engaging in activities further up in the chain (this is known as "forward integration").

The added value of value chain finance

Upgrading in a value chain does not happen automatically. Bottlenecks often occur, especially for small-scale suppliers. Value chain finance can help overcome some of these bottlenecks and smoothen or upscale a chain, so increasing the chain's competitiveness. But its power remains limited if it is not accompanied with market links, business skills, information flows, the right policy framework, enforcement mechanisms, and so on. Value chain finance is not the ultimate solution for a problem in a chain and it does not replace existing financial services. It makes new sources and finance available for rural entrepreneurs, and through this contributes to the empowerment of the poor.

Value chain finance and microfinance

Value chain finance both complements and goes beyond microfinance. An important difference is that value chain finance is tied to actors, transactions and relations in a value chain. Therefore it can reach other actors than microfinance, and at the same time it leaves out actors that (fortunately) microfinance institutions can reach. Many of the cases in this book involve microfinance institutions, which often work together with other financial institutions and farmers' groups. Microfinance can be part of value chain finance, but needs to be complemented with other types of financial services that can address varied needs in the chain.

Key messages

The cases have contributed to many new insights on what value chain finance entails. In the final section of this book, we offer some key messages for different actors involved in value chain finance.

Farmers, traders and other value chain actors

If you work together, you will improve your bankability. The relationship between different chain actors becomes an alternative for the hard collateral that financial institutions normally demand. The combination of more tailor-made financial services with non-financial services builds the capacity of the different chain actors and the chain as a whole.

Banks and other financial institutions

Value chain finance is a serious market opportunity even in rural areas. It does, however, require another perception of risk management: a shift from credit risk to performance risk. It also requires information on the value chain you are considering investing in. Knowledge on how a value chain works opens new opportunities for finance, sometimes demanding new kinds of financial products and new kinds of partnerships. It is a way of increasing your number of clients and expanding your portfolio.

Governments

Stop subsidizing credit; focus on enabling conditions. Financial services should not be subsidized. This disrupts the market and creates a culture of "free money". Governments have a role to play in ensuring that the right infrastructure and policy frameworks are in place to make the value chain operate smoothly. Although ideally governments should not be involved as actors in value chain finance, in some countries and some sectors they do play a role, which cannot be neglected.

Donors and development agencies

Use mid-chain lead firms to reach producers, but beware of creating dependencies. It is challenging to link producers to lead firms in a way that contributes to their empowerment. If this is not an option, you could support the involved producers to move away from the lead firm and look for alternative sources of income. Donors and development agencies can also play other roles in value chain finance, such as stimulating product development, providing non-financial services and building capacity.

Research

Look into models of "chains of value chain finance", and at statistical credit ratings based on certification and other forms of chain integration. Look at best practices and try to find ways to replicate successes, while taking into account contextual differences. Make research outcomes available for the different actors in the chain and for the financial institutions involved.

8

Resources

THIS CHAPTER CONTAINS LINKS to third-party websites and organizations devoted to finance for farmers and other rural micro-entrepreneurs. It also contains references cited in the text, suggestions for further reading on value chain finance, and the contact details of the participants who contributed to this book.

We have replaced long web addresses with a shorter equivalent – the tinyurl addresses in the list below.

Organizations and websites

Agri-ProFocus

apf-finance.ning.com

This online platform facilitates exchange and joint action between Agri-ProFocus' 29 member organizations in the Netherlands and their partner organizations working on finance for farmer entrepreneurship. Agri-ProFocus links value chain development with acces to financial services and sustainable food production. Activities focus on Africa and are coordinated with the MicroNed rural finance group.

CERISE

www.cerise-microfinance.org

CERISE (Comité d'Echanges, de Réflexion et d'Information sur les systèmes d'Epargne -crédit) is an initiative of four French organizations working with microfinance in countries in the South.

CGAP

www.microfinancegateway.org

CGAP is a web-based resource centre for microfinance, with news, announcements, features, a library and key areas.

FAO

tinyurl.com/yhspeep

The value chain finance section of the Food and Agriculture Organization of the United Nations.

FAST International

www.fastinternational.org

The Finance Alliance for Sustainable Trade (FAST) is a global, member-driven, non-profit association of lenders and producers who bring sustainable products to market. FAST aims to increase the number of producers in developing nations who can successfully access quality trade finance, tailored to their business needs, as they enter sustainable markets.

IFC

www.ifc.org/microfinance

The International Finance Corporation, part of the World Bank Group, provides advisory services and direct and indirect investment services to the microfinance sector. Its focus is on creating and supporting commercially viable microfinance institutions that can attract the private capital needed to scale up and respond to unmet demand. It demonstrates the business case for commercial microfinance and promotes it as an asset class to private institutional investors.

Microlinks

www.microlinks.org

Supported by the US Agency for International Development, Microlinks produces a large number of research reports on finance (value chain finance, rural finance, microfinance).

MicroNed

www.micro-ned.nl

MicroNed was created in 2006, as a network of the Dutch Development Finance Organisations Cordaid, Hivos, ICCO and Oxfam Novib, to cooperate on a structural basis with respect to the establishment of a specialised sector approach for microfinance. In 2008 Rabobank Foundation joined MicroNed. The MicroNed rural finance group coordinates activities with Agri-ProFocus.

Rural Finance Learning Centre

www.ruralfinance.org

The Rural Finance Learning Centre aims to assist organizations in developing countries to build their capacity to deliver improved financial services which meet the needs of rural households and businesses.

Shared Interest

www.shared-interest.com

Shared Interest is a co-operative lending society that provides fair and just financial services. Part of the fair trade movement, it works extensively with community-based businesses in Africa and other continents. It provides finance up front to producers, often via their buyers, to enable them to buy raw materials, tools and the other things they need, when they need them. It also offers longer-term loans and other credit facilities to support the development of fair trade businesses.

SEEP

www.seepnetwork.org

The Small Enterprise Education and Promotion (SEEP) Network connects microenterprise practitioners from around the world to develop practical guidance and tools, build capacity, and help set standards to advance their common vision: a sustainable income in every household. SEEP produces materials on a wide range of topics, including value chain finance.

Synergies

www.itcilo.org/synergies

A platform on current applied research and practical implementation of linking financial services to business services that contribute to pro-poor enterprise development, income and job creation. Managed by the International Training Centre of the International Labour Organisation.

References and further reading

Allen, H. and M. Staehle. 2007. Village savings and loan associations (VSLAs). Programme guide. Field operations manual. *www.vsla.net*

Athmer, G. 2008. Rural finance study. An inventory of policies, practices and challenges regarding support to rural (micro)finance among Netherlands' MicroNed members. MicroNed, ICCO and Terrafina Microfinance, Netherlands. *tinyurl.com/yd4dkre*

Bernhardt, J., S. Azar, and J. Klaehn. 2009. Integrated financing for value chains. World Council of Credit Unions, Madison, WI. *tinyurl.com/ylgn6yo*

Charitonenko, S., L. Heron, G. Chalmers, B. Lennon and M. Miller. 2005. Value chain finance. RAFI notes 2. Rural and Agricultural Finance Initiative, United States Agency for International Development. *tinyurl.com/yaoe8fk*

Christen, R.P. and D. Pearce. 2005. Managing risks and designing products for agricultural microfinance: Features of an emerging model. CGAP Occasional Paper 11. *tinyurl.com/yedvrdp*

de Klerk, T. 2008. The rural financial landscape: A practitioner's guide. Agrodok 49. Agromisa Foundation and CTA, Wageningen. *tinyurl.com/yfbwzxf*

Dichter, T. 2007. A second look at microfinance. The sequence of growth and credit in economic history. Development Briefing Paper 1. Cato Institute, Washington, DC. *tinyurl.com/dgq75j*

Dichter, T., and M. Harper (eds). 2007. What's wrong with microfinance? PA Publications, Rugby, and Rawat Press, Jaipur.

Dutta, S. 2001. The livelihood resource manual. New Economic Foundation, India.

Enterprise Development and Microfinance 19(4), October 2008. Special issue on value chain finance. *tinyurl.com/yjm6xzq*

FAO and GTZ. 2004. Agricultural production lending. A toolkit for loan officers and loan portfolio managers. FAO and GTZ, Eschborn 2004. *tinyurl.com/yfoafk7*

Fries, B, and B. Akin. 2004. Value chains and their significance for addressing the rural finance challenge. MicroREPORT 20. Accelerated Microenterprise Advancement Project, United States Agency for International Development. *tinyurl.com/ybey9ox*

Fries, B. 2005. The value chain framework and rural finance. PowerPoint presentation. *tinyurl.com/yl2sp4b*

Galvez, E. 2006. Financiación de la comercialización agrícola en América Latina. FAO. Servicios de Gestión, Comercialización y Finanzas Agrícolas (AGSF). FAO. Rome. *tinyurl.com/yj8dfnd*

González, A. and R. Rosenberg. 2006. The state of microfinance: Outreach, profitability, and poverty. Findings from a database of 2,600 microfinance institutions Consultative Group to Assist the Poor, Washington, DC.

Habyalimana, S. 2007. Access to finance through value chains: Lessons from Rwanda. *tinyurl.com/yb37jtm*

Harper, M. 2009. Inclusive value chains in India: Linking the smallest producers to modern markets. Global Scientific Publishers, Singapore. (Available from *ivc-india@hotmail.com*)

Harper, M., D.S.K. Rao and A.K. Sahu. 2008. Development, divinity and dharma: The role of religion in development institutions and microfinance. Practical Action Publications, Rugby. *tinyurl.com/yfoa4as*

International Year of Microcredit, 2005. *tinyurl.com/ydkcomx*

Kauffmann, C. 2005. Financing SMEs in Africa. OECD Policy Insights 7. OECD Development Centre, Paris. *tinyurl.com/y95p2c6*

KIT and IIRR. 2008. Trading up: Building cooperation between farmers and traders in Africa. Royal Tropical Institute, Amsterdam; and International Institute of Rural Reconstruction, Nairobi. *tinyurl.com/4jqyes*

KIT, Faida MaLi and IIRR. 2006. Chain empowerment: Supporting African farmers to develop markets. Royal Tropical Institute, Amsterdam; Faida MaLi, Arusha; and International Institute of Rural Reconstruction, Nairobi. *tinyurl.com/yjjuqwt*

KIVA. 2009. About microfinance. *tinyurl.com/2cx6fy*

Miller, C. no date. Rural finance course. Calvin Miller, FAO, IDEAS and the Giordano dell' Amore Foundation in collaboration with the University of Bergamo, Italy.

Nagarjan, G. and R.L. Meyer. 2005. Rural finance: Recent advances and emerging lessons, debates and opportunities. Reformatted version of Working Paper AEDE-WP-0041-05, Dept of Agricultural, Environmental, and Development Economics, Ohio State University, Columbus, Ohio. *tinyurl.com/yg9jk6l*

Nair, A., R. Kloeppinger-Todd, and A. Mulder. 2004. Leasing: An underutilized tool in rural finance. World Bank Agricultural and Rural Development Discussion Paper 7. *tinyurl.com/yzptx6d*

Neven, D. 2009. Finance in value chain analysis: A synthesis paper. microREPORT 132, United States Agency for International Development. *tinyurl.com/y8pcrkv*

OECD. 2006. The SME finance gap. Vol. 1: Theory and evidence. Organisation for Economic Co-operation and Development, Paris. *tinyurl.com/ybfj3ay*

Pagura, M. 2004. Mapping rural finance products, delivery models and linkages. PowerPoint presentation. *tinyurl.com/yl2sp4b*

Quirós, R. (ed.) 2007. Agricultural value chain finance. Summary of the conference "Agricultural value chain finance", Costa Rica, 16–18 May 2006. Academia de Centroamérica, San José; FAO; RUTA; Serfirural. *tinyurl.com/yc3zgtd*

Seibel, H. 2005. Rural finance innovations. Topics and case studies. World Bank Publications. Agricultural and Rural Development Department, Washington.

Seibel, H.D. 2005. Finance for small-scale commodity processing: From micro to meso-finance. Proceedings of an international workshop in Khartoum, Sudan, 9–11 November 2003. Common Fund for Commodities, Technical Paper 32. *tinyurl.com/ygukbwe*

Shepherd, A. 2004. Financing agricultural marketing: The Asian experience. Agricultural Management, Marketing and Finance Occasional Paper 2. Food and Agriculture Organization of the United Nations, Rome. *tinyurl.com/yevahor*

The Economist. 18 July 2009. The effects of microcredit. *tinyurl.com/l4858r*

UNCTAD. 2004. Financing commodity-based trade and development: Innovative agriculture financing mechanisms. Report prepared by the UNCTAD secretariat for the Expert Meeting on Financing Commodity-Based Trade and Development: Innovative Financing Mechanisms. Geneva, 16–17 Nov 2004. TD/B/COM.1/EM.24/2, 3 Sep 2004. United Nations Conference on Trade and Development, Geneva. *tinyurl.com/y8azepg*

USAID. 2005. Value chain finance. RAFI Notes 2, United States Agency for International Development, Washington, DC. *tinyurl.com/3pbba4*

WOCCU. 2003. A technical guide to rural finance: Exploring products. WOCCU Technical Guide 3. World Council of Credit Unions, Madison, WI. *tinyurl.com/yg5b2j6*

World Bank. 2005. Rural finance innovations: Topics and case studies. Report 32726-GLB World Bank, Agriculture and Rural Development Department. *tinyurl.com/yaey7yq*

Yago, G., D. Ropveda, and J.M. White. 2007. Transatlantic innovations in affordable capital for small- and medium-sized enterprises. Prospects for market-based development finance. German Marshall Fund of the United States and Milken Institute. *www.tinyurl.com/3zworo*

Yollin, P. 2007. Microcredit movement tackling poverty one tiny loan at a time. *San Francisco Chronicle*, 30 Sep 2007. *tinyurl.com/yfhzrvf*

Yunus, M. 2007. Creating a world without poverty. Social business and the future of capitalism. Public Affairs, New York.

Participants' profiles

This section gives the names, contact details and brief biographies of participants at the Nairobi writeshop, where the draft of this book was prepared.

Sangeeta Agasty

Manager, Foundation for MSME Clusters

USO House, USO Road, Off Shaheed Jeet Singh Marg, 6, Special Institutional Area, New Delhi 110 067, India

Tel. +91 9903168836, fax +91 11 41688589, 41688590, email sangeeta.agasty@gmail.com

Sangeeta holds a master's degree in rural management and an MS in economics. Her experience includes implementing and managing livelihood projects, the development of clusters of micro-, small and medium enterprises (MSME), action research, training and capacity building, monitoring and evaluation of development projects and participatory development approaches. She worked with the United Nations Industrial Development Organization (UNIDO) as national expert under its cluster development programme, and currently works with the Foundation for MSME Clusters. She has developed training modules on financial literacy and network management for micro-enterprises, and has studied livelihood, finance and social research methods.

Navil Agramont Akiyama

Commercialization and market specialist, Save the Children Bolivia

PO Box 15120, La Paz, Bolivia

Tel. +591 2 2481606, +591 2 2481615, mobile +591 701 12136, fax +591 2 2115856, email nagramont@savechildren.org.bo, nagramont@hotmail.com

Navil has a degree in economics, and a master's in business administration, postgraduate training in project preparation and evaluation, and a diploma in marketing and financial engineering. He has worked as a marketing specialist, university teacher and consultant for various organizations and rural enterprises. At Save the Children, he focuses on developing and funding value chains.

Tesfaye Befekadu

General manager, Harbu Microfinance Institution Share Company

PO Box 13640, Addis Ababa, Ethiopia, fax +251 11 618 0654, email harbumfi@ethionet.et, harbumfi@yahoo.com, tesfayebef@yahoo.com

Tesfaye is a graduate in accounting with experience mainly in development-oriented organizations. For the last 12 years he has worked in various microfinance institutions as a specialist and manager. He provides training and consulting services to organizations and individuals in establishing microfinance institutions.

María Renée Bejarano Vega

General coordinator, SEED Programme, Vice Ministry of Equal Opportunities, Bolivia

Tel. +591 2 2119371, +591 2 2110492, fax +591 2 2119758, email renee.bejarano@gmail.com, mr.bejarano@semilla.pnud.bo

María Renée holds a master in business administration and has extensive experience in rural development management, working with farmer organizations, micro- and small enterprises, non-government organizations and local governments. Until 2009 she was executive director of Pro-Rural, a rural development NGO in Bolivia (*www.prorural.org.bo*). She now coordinates a $9 million fund to increase the economic and social opportunities of rural women through financial services, implemented by the United Nations Development Programme and funded by the Spanish Agency for International Cooperation for Development.

Isaac Bekalo Bateno

President and Regional Director for Africa, International Institute of Rural Reconstruction (IIRR)

PO Box 66873, Nairobi 00800, Kenya

Tel. +254 20 4442610, 4440991, fax +254 20 4448814, email isaac.bekalo@iirr.org, website www.iirr.org

Isaac holds a PhD in organizational development and planning. His over 30 years experience includes teaching, community development, NGO training, curriculum design and organizational development. He provides technical services on strategic planning, business plan development, organizational diagnoses, participatory monitoring and evaluation, project design and proposal writing, and facilitation of global forums. He specializes in participatory development approaches and organizational development.

Manuel Antonio Bermúdez Vado

Business manager, Fondo de Desarrollo Local (Fund for Local Development, FDL)

A-339, Managua, Nicaragua

Tel. +505 2774245, 2707371, 2707338, fax +505 2774245, email mbermudez@fdl.org.ni, website www.fdl.org.ni

Manuel holds a master's in business administration with a focus on finance. He has been working in rural microfinance for 15 years, the last 10 years as a business manager. As part of his work with FDL, he advises other microfinance institutions in Central America.

Nyotumba Bonaventure

Artist/graphic designer

Email nyotsz@yahoo.com, website www.developmentart.com/artists.htm

Bonaventure is a freelance designer-cum-artist based in Nairobi. He has a diploma in fine art. He has worked as a designer/painter for Bellerive Foundation, CARE-Kenya, Rainbow magazine, Jacaranda Designs, Don Bosco, Jericho Church and the International Institute of Rural Reconstruction. He specializes in fine and graphic art, product design and desktop publishing.

Harry Clemens

Programme officer, financial services and enterprise development, HIVOS

Raamweg 16, PO Box 85565, 2508 CG, The Hague, Netherlands

Tel. +31 70 376 5500, fax +31 70 362 4600, email h.clemens@hivos.nl

Harry is a development economist who graduated from the Free University, Amsterdam. He has more than 20 years of experience in development projects and consultancy services, with emphasis on rural development and financial services. He has also worked on social development, agricultural economics, food security and markets.

Tarekegn Garomsa

Programme officer for planning, monitoring and evaluation, Facilitators for Change Ethiopia (FCE)

PO Box 24199, code 1000 or 14180, Addis Ababa, Ethiopia

Tel. +251 116621438, 6517535, mobile +251 913 289151, fax +251 11 618 0654, email fce@ethionet.et, taregarom@yahoo.com, website www.fce-eth.org

Tarekegn holds a postgraduate diploma in rural development management (specialized in small and medium enterprises and entrepreneurship development) from the International Centre for Entrepreneurship and Career Development, India. He obtained his BA in sociology and social administration from Addis Ababa University, Ethiopia. He has practical grassroots experience in rural development (working with farmers' marketing organizations and women's self-help groups), value chain development, project coordination, participatory planning, monitoring and evaluation, and rural project design and implementation.

Straton Habyalimana

Economic development adviser, SNV Netherlands Development Organisation

PO Box 1049, Kigali, Boulevard d l'Umwganda 1, Rwanda

Tel. +250 78883 5771, +250 575619, fax +250 574671, email shabyarimana@snvworld.org

Straton holds an MBA in project management and qualifications in economics, microfinance and community economic development as well as public finance. His 20 years of experience include teaching, management of public finance departments, technical support to local organizations (institutional and organizational development, strategic and business planning, training and networking), and microfinance. He specializes in market research and product development for microfinance institutions, and business development services for small and medium enterprises.

Malcolm Harper

The Old Farm House, 37 Filgrave, Buckinghamshire, MK16 9ET, United Kingdom

Tel. +44 1234 711764, email malcolm.harper@btinternet.com

Malcolm was educated at Oxford, Harvard and Nairobi. He worked in marketing in England, and then taught at the University of Nairobi. He was professor of enterprise development at Cranfield, and since 1995 he has worked independently, mainly in India. He has also worked on poverty issues in many other parts of the developing world. He has published on self-employment, enterprise development, microfinance and livelihoods. He was chairman of BASIX Finance for 10 years, and is chairman of Micro-Credit Ratings International Ltd (M-CRIL). He was the founding editor of *Small Enterprise Development*, and is a director and trustee of Homeless International, EDA (UK) Ltd, APT Enterprise Development and Practical Action Publications in the United Kingdom.

Joseph Kabundi

General director, Caisse des Affaires Financières Isonga (CAF Isonga)

PO Box 140, Gitarama, Rwanda

Tel. +250 788302039, 788524537, fax +250 252 562230, email jkabundi@yahoo.fr, caf_isonga@yahoo.fr

Joseph has a qualification in development studies. His experience includes school management, social services to street children, and financial and microfinance management. He is the founder of CAF Isonga, a microfinance institution, of which he was managing director from 2004 to 2006 and chief executive officer from then on.

Paul Karaimu

Knowledge and documentation coordinator, International Institute of Rural Reconstruction

PO Box 66873, Nairobi 00800, Kenya

Tel. +254 20 4442610, 4440991, fax +254 20 4448814, email paul.karaimu@iirr.org

Paul holds a BA in communications and a certificate in electronic publishing. He has experience in writing, editing and computer-aided design. He previously worked as an editor for a book publisher and as a writer/editor for a magazine publisher in Kenya.

Maurits de Koning

International Sales Manager, Wallaboo

Email hjmdekoning@hotmail.com

Maurits is a business economist and cultural anthropologist specializing in sustainable economic development. His expertise lies in business development services, financial services (microfinance, savings and credit unions) and chain development. Until 2009 he was an adviser in sustainable economic development at the Royal Tropical Institute (KIT).

Anna Laven

Adviser in sustainable economic development, Royal Tropical Institute (KIT)

PO Box 95001, 1090 HA Amsterdam, Netherlands

Tel. +31 205688207, fax +31 205688444, email a.laven@kit.nl

Anna holds an MA in political science and environmental management, and is finalizing her PhD on governance and upgrading in value chains. Her experience includes research, teaching, multi-stakeholder workshops, proposal writing and policy making. She specializes in sustainable development issues and value chain development.

Billy O Mathu

Project officer, Kenya Gatsby Trust

PO Box 16052, Nairobi 00610, Kenya

Tel. +254 20 2720711, 2720703, fax +254 20 2721707, email bmathu@kenyagatsby.org, bllymathu@yahoo.co.uk, website www.kenyagatsby.org

Billy is a certified public accountant with experience in financial modelling and analysis, project monitoring, business training and advisory services. He is involved in financial product development and promotion of factoring products. He is currently studying for an MBA with a speciality in strategic management.

Oliver Mundy

Student, Philipps University of Marburg

Müllenberg 5a, 51515 Kürten, Germany

Tel. +49 2268 801691, fax +49 2268 801692, email oliver@mamud.com

Oliver is studying for his bachelor's degree in geography, and expects to graduate in 2010 after a year's study at the Sorbonne in Paris. He recently completed 11 months of community service with Growing Nations, an NGO in Lesotho, where he grew maize and beans for AIDS orphans using conservation agriculture techniques, and taught computer skills to the NGO staff and local young people.

Paul Mundy

Independent consultant in development communication

Müllenberg 5a, 51515 Kürten, Germany

Tel. +49 2268 801691, fax +49 2268 801692, email paul@mamud.com, website www.mamud.com

Paul is a British consultant in development communication. He holds a PhD in journalism and mass communications from the University of Wisconsin-Madison. He specializes in easy-to-understand extension materials, developed through intensive writeshops like the one used to produce this book. He also provides consultancy services in various aspects of development communication. He has worked extensively in Africa, Asia, Latin America and the Caribbean.

Paul Maundu Mwilu

Microfinance officer, K-Rep Development Agency

PO Box 10528, Nairobi 00100, Kenya

Tel. +254 721 902 731, email maundumwilu@yahoo.com

Paul holds a degree in sociology and economics from Egerton University, Kenya, and has received a postgraduate advanced certificate in marketing management. He has worked in microfinance, specializing in new product development for K-Rep Development Agency for financing values chains in poultry, dairy, cotton and honey.

Joseph Muriithi Ndegwa

Coordinator/senior training and development consultant, Future Focus Development

PO Box 328, Naivasha 20117, Kenya

Tel. +254 733858953, 726275442, email jmundegwa@yahoo.com, fdevelopment7@gmail.com

Joseph holds a BA in human and social studies (specializing in development studies) from the University of South Africa, Pretoria, diplomas in agriculture and community-based development, and a certificate in agricultural extension. He has 26 years of experience in public and private organizations, managing integrated projects involving capacity building and training. He has provided consultancy services in project design, implementation, capacity building, monitoring and evaluation. Until 2009, he was agricultural coordinator with the economic programme of Nakuru Region Inter-Diocesan Christian Community Services (NRIDCCS). He is accredited as a business development service provider by the Kenya Dairy Board/Ministry of Livestock Development. He also lectures part-time to diploma students at the Kenya Institute of Management.

Joseph Njoroge Ndirangu

Farmer and secretary, Kwieria Tangi Women Group

PO Box 42, Rumuruti, Kenya

Tel. +254 725692764

Joseph graduated from Kisumu Technical High School and has gathered a wide range of experience in general construction. He is specialized in carpentry, joinery, metalwork and masonry, as well as being experienced in farm management. He has earned a certificate in facilitation from a farmer field school. He is involved in several farmer groups as a mobilizer and group member, and helps manage a community-based organization, the Kwieria Tangi Women Group, that constructs water tanks for its members and has gone into chilli production to earn money.

Callixte Niyonsaba

Farmer and member, COPRORIZ

c/o PO Box 140, Gitarama, Rwanda

Tel. +250 788681267

Callixte is a farmer who started growing rice in 2003 as an add-on to his usual crops, maize and sorghum. He is a member of the Mukunguli rice producers' cooperative, COPRORIZ.

Janet Nyaoro

Regional training manager, International Institute of Rural Reconstruction, Africa Regional Center

PO Box 66873, Nairobi 00800, Kenya

Tel. +254 722 230168, 20 444 2610, fax +254 20 444 8814, email janet.nyaoro@iirr.org, website www.iirr.org

Janet is a trainer/facilitator with 9 years of experience in training in the for-profit and humanitarian sectors. At IIRR, she facilitates training on participatory monitoring and evaluation, community-managed disaster risk reduction and value chain development, and also facilitates workshops and writeshops. She is also involved in organizational development, baseline surveys and capacity assessment for national and regional programmes. She holds a BEd in economics and geography and a postgraduate diploma in human resource management.

Marcio Oblitas Fernández

General manager, Sartawi Foundation

PO Box 8050, La Paz, Bolivia

Tel. +591 2 2481000 or 591 2 2482000, mobile +591 70195195, fax +591 2 2487000, email marcio.oblitas@sartawi.org, marcio_oblitas@hotmail.com, website www.sartawi.org

A civil engineer by training, Marcio has a postgraduate degree in microfinance and specializations in social management and the management of projects. He has experience in rural development and financial analysis with several programmes funded by the Inter-American Development Bank, the World Bank, USAID, GTZ and others. In addition, he serves as vice-president of FINRURAL (Bolivian Network of Financial Institutions for Rural Development), as an assembly member of the Sartawi Foundation, and assembly president of FOCADES (an NGO that provides technical assistance to micro-entrepreneurs).

Aileen Ogolla

Regional communications manager, International Institute of Rural Reconstruction

PO Box 66873, Nairobi 00800, Kenya

Tel. +254 722 230168, 20 444 2610, fax +254 20 444 8814, email aileen.ogolla@iirr.org, website www.iirr.org

Aileen holds an MA in communication. She has experience in various activities in communication, training facilitation and programme coordination.

Alfred Ombati

Artist/graphic designer

PO Box 64427, Nairobi 00600, Kenya

Tel. +254 723 350628, 721 420806, email aholiabsart@yahoo.com

Alfred is a freelance artist who has been involved in publication projects with various NGOs, including the International Institute of Rural Reconstruction, Médecins Sans Frontières, and the African Technology Policy Studies Network; publishing companies including Cover Concept, Oxford University Press, Sahel Book Publishing Co., and the Ministry of Education; and advertising agencies such as Lowe Scanad, McCann Erickson, and Quite Bright Films. He also does fine arts (painting), murals and printing.

Beeranna Palyam

Zonal manager, Reliance Dairy Foods Limited

Plot 405, Ashtalaxmi Apartments, Station Road, Mahabubnagar 509001, Andhra Pradesh, India

Tel. +91 99 897 73799, email beeranna_palyam@rediffmail.com

Beeranna earned his bachelor's degree in veterinary science in 1977. He has 31 years of experience in dairy development, project preparation and institution building, including buffalo breeding (for the Indian Council of Agricultural Research), animal management, the formation of cooperative societies, and milk procurement. He was also involved in implementing a training and employment programme in Karnataka. As zonal manager with Reliance Dairy Foods, he manages cooperatives and helps integrate them with Reliance's corporate business activities.

Niranjan Pattni

Managing director, BioRe Tanzania Limited

PO Box 49, Meatu District, Mwamishali-Shinyanga, Tanzania

Tel. +255 28 2795090, mobile +255 784860854, fax +255 28 2795027, email niranjan.pattni@biore-tanzania.com

Niranjan is the managing director of a company that grows organic cotton, and has worked there since the company's founding. After graduating from high school, he worked for two years as a teacher in a primary school, before beginning work in a textile mill. He retired as general manager from the mill in 1996. Niranjan has a lot of experience in the textile industry and in cotton-growing with smallholder farmers in Tanzania.

Lucian Peppelenbos

Learning manager, Dutch Sustainable Trade Initiative

Waterstraat 47, Postbus 48, 3500 AA Utrecht, Netherlands

Tel. +31 30 2363 488 / 485, fax +31 30 2312 804, email peppelenbos@dutchsustainabletrade.com

Lucian holds a PhD in agricultural sciences from Wageningen University, where he specialized in value chain management. Until 2009 he was a senior adviser at the Royal Tropical Institute (KIT), where he worked on pro-poor business development in Africa, Latin America, and India. Before this, he worked in Chile as a management consultant for export agribusiness firms, farmer cooperatives and international agencies. Lucian has recently joined the Dutch Sustainable Trade Initiative, where he works on the mainstreaming of sustainability in international commodity chains.

Nagabhushana Clospet Satyanarayana Rao

Senior consultant, Catalyst Group

19, 1st Main, 1st Cross, Ashwathnagar, RMV 2nd Stage, Bangalore 560 094, India

Tel. +91 9343829737, +91 80 23419616. 80 23417714, email nagabhushana@cms-india.org, website www.catalysts.org, www.swasti.org

Nagabhushana holds postgraduate degrees in statistics and economics and has some 30 years of experience in the development sector. His experience includes conducting evaluation studies, value chain analyses, baseline surveys, training, database management, and preparing proposals and reports. He has extensive experience in working on assignments funded by the World Bank, Danida, DfID, and GTZ, as well as government and non-government organizations in India. He has been a mission member for Danida and GTZ to review institutional aspects of watershed projects in India.

Luisa Santur Alberca

Microenterprise services manager, Fomento de la Vida (Fovida)

Av. Javier Prado Oeste 109, Magdalena del Mar, Lima 17, Peru

Tel. +51 1 2617548, fax +51 1 4610106, email luisa@fovida.org.pe

Luisa has experience in the design, monitoring and supervising of micro-enterprise development and income-generation projects. She previously worked at the Employment Department in the Ministry of Labour and Social Promotion. She has worked for the NGO Fovida since 1990, where she headed the Department of Economic Development and the Program for Enterprise Affairs, before becoming manager of Micro-enterprise Services. In addition, she has completed her studies for her master's degree in sociology from the Catholic University of Peru.

Alexandra Schütze

Board member, Fondo de Desarrollo Local (FDL)

Edificio FDL, Campus Universitario, Universidad Centro Americana (UCA), Managua, Nicaragua

Tel. +505 2774245, email alexschutze@gmail.com

Alexandra holds an MSc in finance, an MBA, and a BS in sociology. She has been working mainly with financial service providers in capital markets, banking and microfinance. Her experience includes marketing, services and product development, strategic analysis, finance and operations. She has been a member of FDL's Board since 2003. She is now a consultant in financial institution management and small and medium enterprise development.

Sarah Simpson

Communications officer, Royal Tropical Institute (KIT)

PO Box 95001, 1090 HA, Amsterdam, Netherlands

Tel. +31 20 568 8301, +31 20 568 8444, email s.simpson@kit.nl

A Canadian citizen, Sarah is a communications officer for KIT Development Policy and Practice, a department of the Royal Tropical Institute. Previously she worked as an editor and technical writer. She has a background in English literature and a degree in biochemistry.

Amit Kumar Singh

Programme Manager, ACDI/VOCA

ASI, 2/34, Vipul Khand, Gomati Nagar, Lucknow, Uttar Pradesh 226010, India

Tel. +91 9956999684, email asinghks@gmail.com

Amit has experience in conceptualizing, executing and managing socio-economic development projects. He is a specialist in development research, livelihood promotion, cluster development and micro-enterprise promotion. He has published many articles on these subjects in national and international journals. He has worked with the United Nations Industrial Development Organization as a national expert on the sal leaf industry. At the time this book was written he was state coordinator of the Madhya Pradesh Rural Livelihoods Project, *www.mprlp.in*, *www.ruralenterprise.blogspot.com*.

Ephrem Soosai

Chief executive, South Indian Federation of Fishermen Societies (SIFFS)

Karamana, Trivandrum, Kerala 695 002, India

Tel. +91 471 2343711, 2343178, mobile +91 9443143711, fax +91 471 2342053, email ephrem.siffs@gmail.com, website www.siffs.org

After graduating, Ephrem joined a village-level fish marketing society, one of the base-level organizations in the three-tier structure of the South Indian Federation of Fishermen Societies (SIFFS). Now he is the head of the organization, having spent his whole career with the organization and holding many positions within it, starting from the grassroots level. He has attended many training courses in India and internationally on microfinance.

Mendu Srinivasulu

Head, livelihood initiatives, Krishna Bhima Samruddhi Local Area Bank Ltd, BASIX Group

7-4-58/1, P C Towers, Ground floor, Venkateshwara Colony, Mahabubnagar 509002, Andhra Pradesh, India

Tel. +91 93 900 36621, +91 8542 273383/84, email sreenivasulu.m@basixindia.com, mendusreenivas@gmail.com, website www.basixindia.com

Mendu holds a postgraduate diploma in business administration with marketing specialization and also a diploma in agriculture from Acharya NG Ranga Agricultural University. He has 14 years of field experience in agriculture and related development activities, including providing agricultural, business and institutional development services to small and marginal producers of cotton, rice, red gram and milk. He has worked for 8 years in microfinance to link producer groups with formal business development services. Before joining the BASIX group in 2001, he worked with Sterling Tree Magnum (I) Ltd, Rallis India and Ikisan.com in various capacities.

Julia Wagemann

Student, Philipps University of Marburg

Zischendorf 9, 91555 Feuchtwangen, Germany

Tel. +49 7950 9269512, fax +49 7950 1431, email wagemann.julia@gmx.de

Julia is currently studying for her bachelor's degree in geography at the Philipps University of Marburg. She specializes in physical geography and her main interest is in geographical information systems and climatology. In 2010 she will complete her degree with a year's study in Quebec, Canada, after which she is considering studying for a master's degree.